Pseudo-Differential Operators

Theory and Applications

Volume 16

Pseudo-Differential Operators: Theory and Applications is a series of graduate-level textbooks and monographs appealing to students and experts alike. Pseudo-differential operators are understood in a very broad sense and include such topics as harmonic analysis, PDE, geometry, mathematical physics, microlocal analysis, time-frequency analysis, imaging and computations. Modern trends and novel applications in mathematics, natural sciences, medicine, scientific computing, and engineering are highlighted.

Der-Chen Chang • Bert-Wolfgang Schulze

Analysis on Manifolds with Singularities

Der-Chen Chang
Department of Mathematics and Statistics
Georgetown University
Washington, DC, USA

Graduate Institute of Business Administration
College of Management, Fu Jen Catholic
University
New Taipei City, Taiwan

Bert-Wolfgang Schulze
Institut für Mathematik
Universität Potsdam
Potsdam, Germany

ISSN 2297-0355 ISSN 2297-0363 (electronic)
Pseudo-Differential Operators
ISBN 978-3-032-01889-2 ISBN 978-3-032-01890-8 (eBook)
https://doi.org/10.1007/978-3-032-01890-8

Mathematics Subject Classification: Primary 35S35; Secondary 35J70

This book is published under the imprint Birkhäuser, www.birkhauser-science.com by the registered company Springer Nature Switzerland AG
The registered company address is: Gewerbestrasse 11, 6330 Cham, Switzerland

If disposing of this product, please recycle the paper.

Introduction

The relationship between partial differential operators and their symbols is the origin of the calculus of pseudo-differential operators, where the symbols are not necessarily polynomials in the covariables. Basics have been well-known through the works of J.J. Kohn and L. Nirenberg [38], L. Hörmander [33], and many other authors. Standard material can be found in numerous textbooks on this topic. From the very beginning of its development, the pseudo-differential analysis interacted with other fields of mathematics and applications in natural sciences, especially geometry, topology, and physics. In the present exposition, we update the insight into pseudo-differential techniques from the analysis on manifolds with singularities, with new starting points and future possibilities and challenges.

The classical ideas around ellipticity of operators, parametrices, and index on smooth closed manifolds are an inspiration also for the singular analysis. The same is true of other traditional structures occurring in machineries for solving parabolic or hyperbolic problems. The singular geometries in such contexts cause new and partly unexpected problems. Although those are sometimes really hard, the new symbolic structures and iterative ideas that are created so far are very beautiful. Also the remarkable progress during the past decades is an encouragement to participate in these fields. In this exposition we focus on what is necessary for elliptic operators.

The interactions between classical pseudo-differential analysis and other fields of mathematics have attracted mathematicians for a long time. One of the most impressive examples is the index theory of elliptic operators and of boundary value problems (BVPs) on manifolds. I.M. Gelfand, in his article [26], pointed out that the Fredholm index of a Shapiro-Lopatinskij elliptic BVP is a function of the involved principal symbols alone. This led to the question, known as Gelfand's program, how to express the index purely in terms of suitable equivalence classes of symbols. In the subsequent development more aspects and objectives have been included in the discussion, in particular, operator algebras, complex analysis, algebraic topology, homotopy theory, and K-theory. The development culminated in the Atiyah-Singer index theorem, cf. [4]. These achievements first concerned elliptic operators on a smooth closed manifold, where the ellipticity of a classical (pseudo-)differential operator is determined by its homogeneous principal symbol (and topological data of the underlying manifold). Later on the development integrated boundary value problems for operators with the transmission property at the boundary, cf. L. Boutet de Monvel [9]. It has been assumed in this paper that a topological obstruction for the interior symbol vanishes, cf. M.F. Atiyah and R. Bott [2] for the case of differential operators. During this period also operators without the transmission property have been investigated, see the theory of pseudo-differential BVPs of M.I. Vishik and G.I. Eskin [81, 82], and G.I. Eskin [24], based on a higher-dimensional analogue of Wiener-Hopf techniques. Also questions around parametrices and the index have been considered in these works.

A "well-organized" pseudo-differential theory should be able to express the parametrices of elliptic operators within a corresponding operator algebra. This is the case in "standard" pseudo-differential operators on a closed smooth manifold. Then the ellipticity of the corresponding operator, i.e., bijectivity or non-vanishing of the homogeneous principal symbol, is equivalent to its Fredholm property in standard Sobolev spaces. If the manifold is non-compact and smooth, ellipticity and parametrices can be formulated as well. However, the nature of "coefficients" of operators and also the kind of smoothing operators close to the non-compact ends of the manifold are very essential for Fredholmness and index in suitable analogues of Sobolev spaces. The difference between different notions of ellipticity and index theory can be illustrated by comparing different compactifications of the respective manifold. For instance, if M is a smooth compact manifold, $v \in M$ a single point, then $M \setminus \{v\}$ can be first compactified to $M_0 := M$ itself. However, when v is regarded as an embedded conical singularity we write M_1 rather than M. Another compactification M_2 of $M \setminus \{v\}$ consists of a manifold with smooth boundary, where ∂M_2 is a sphere of dimension $\dim M - 1$. While for M_0 we have the usual pseudo-differential calculus on M in the smooth sense across v, on M_1 there is the calculus on $M_1 \setminus \{v\}$ with M_1 as a manifold with conical singularity v, which is rather different from the smooth one close to v Moreover, M_2 admits the completely different calculus of BVPs. We then have even different choices of non-equivalent theories, e.g., BVPs with or without the transmission property at the boundary. The spaces M_0, M_1, and M_2 are examples of stratified spaces, and adequate pseudo-differential structures are determined by corresponding symbolic hierarchies, the components of which are contributed by the involved strata. We can also compactify $M \setminus \{v\}$ in such a way that the boundary is a manifold with edges or corners, i.e., stratified again. Then for operators and their ellipticity we need symbolic hierarchies of more than 2 components.

In general it is to be expected that parametrices of partial differential operators on manifolds with corner singularities in such a sense, including with various types of non-compact "exits to infinity," are pseudo-differential in the corresponding singular framework. This is also an ongoing area of active research. The motivation lies in numerous applications and new challenges, see [31, 36] or [70].

We focus here on new approaches from the analysis of pseudo-differential operators, in particular, for spaces with regular conical and edge singularities, and also singularities of higher order. The investigation of conical singularities has a long history of the singular analysis and the variety of different aspects of the analysis that employs Mellin techniques, cf. M. Dauge [19], G.I. Eskin [24], A.I. Komech [39], V.A. Kondratyev [40], Pham The Lai [43], B.A. Plamenevskij [53–55], V.S. Rabinovich [56, 57], G. Cheeger [17], R. Melrose [49], S. Rempel, and B.-W. Schulze [58, 60].

The idea of this book is to pick some crucial parts of recent inventions in singular analysis which contribute to the indicated pseudo-differential structures and to establish new insight to building up singular operators in corresponding operator algebras with symbolic structures.

The authors are very grateful to Professors D.H. Phong and P. Greiner for their valuable suggestions and friendship. We thank Mahdi Hedayat Mahmudi, Sara Khalil, Peter Grabs, and many former members of the research group "Partial Differential Equations and Complex Analysis" of the University of Potsdam for cooperation and valuable discussions on details of this exposition. The first author wants to thank the Georgetown University, especially Professor Robert Groves for long-term support. The authors also want to thank Birkhäuser/Springer, especially Mazlum, for constant generous encouragement to finishing this manuscript.

Contents

Pseudo-differential Operators: Basics and New Techniques

1

1.1 Elements of the Local Calculus

1.1.1 Motivation and Examples

One of powerful tools to study partial differential equations is Fourier transform. Let us start with a well-known example. Consider the equation

$$A(u) = (1 - \Delta)u = \left(1 - \sum_{k=1}^{n} \frac{\partial^2}{\partial x_k^2}\right)u = f \tag{1.1}$$

with Δ being the Laplacian in $\mathbb{R}^n$ where $f \in C_0^\infty(\mathbb{R}^n)$. It is easy to see that the Fourier transform of $\frac{\partial u}{\partial x_j}$ is

$$F\left(\frac{\partial u}{\partial x_j}\right)(\xi) = \left(\widehat{\frac{\partial u}{\partial x_j}}\right)(\xi) = \int_{\mathbb{R}^n} e^{-ix\cdot\xi}\frac{\partial u}{\partial x_j}(x)dx = -i\xi_j\widehat{u}(\xi). \tag{1.2}$$

Therefore,

$$\left(1 + \sum_{k=1}^{n} \xi_k^2\right)\widehat{u}(\xi) = (1 + |\xi|^2)\widehat{u}(\xi) = \widehat{f}(\xi).$$

It follows that

$$u(x) = \frac{1}{(2\pi)^n} \int_{\mathbb{R}^n} e^{ix\cdot\xi}\widehat{u}(\xi)d\xi = \frac{1}{(2\pi)^n} \int_{\mathbb{R}^n} e^{ix\cdot\xi}(1 + |\xi|^2)^{-1}\widehat{f}(\xi)\,d\xi.$$

© Springer Nature Switzerland AG 2025
D.-C. Chang, B.-W. Schulze, *Analysis on Manifolds with Singularities,*
Pseudo-Differential Operators 16, https://doi.org/10.1007/978-3-032-01890-8_1

Now let us move to partial differential operators with variable coefficients. Consider the equation

$$A(x, D)u = \sum_{|\alpha| \leq m} a_\alpha(x) D^\alpha u = f. \tag{1.3}$$

It follows that

$$A(x, D)u(x) = \int_{\mathbb{R}^n} e^{ix \cdot \xi} a(x, \xi) \widehat{u}(\xi) \, đ\xi$$

where $đ\xi = \frac{1}{(2\pi)^n} d\xi$ and $a(x, \xi) = \sum_{|\alpha| \leq m} a_\alpha(x) \xi^\alpha$.

Freezing the coefficients at x_0, can we use $A(x_0, D)u(x)$ to approximate the solution of the Eq. (1.3)? In general, this is not true. Let

$$\sigma_\psi(A)(x, \xi) = \sum_{|\alpha| = m} a_\alpha(x) \xi^\alpha, \qquad x \in \mathbb{R}^n, \qquad \xi \in \mathbb{R}^n \setminus \{0\}$$

the *homogeneous principal symbol* of $a(x, \xi)$. We say that A is *elliptic* if $\sigma_\psi(A)$ does not vanish for all $(x, \xi) \in T^*\mathbb{R}^n \setminus 0$. As usual, T^*X denotes the cotangent bundle of the manifold X and 0 indicates the zero section, locally represented by $\xi = 0$. In this case, an approximate solution of (1.3) should be obtained in the form

$$u(x) \approx \int_{\mathbb{R}^n} e^{ix \cdot \xi} p(x, \xi) \widehat{f}(\xi) đ\xi$$

for a suitable function $p(x, \xi)$ that is not necessarily a polynomial in ξ. In general, an operator that is able to express solvability of elliptic equations is expected to be of the form

$$\mathrm{Op}(a)(f)(x) = \int_{\mathbb{R}^n} e^{ix \cdot \xi} a(x, \xi) \widehat{f}(\xi) \, đ\xi, \qquad f \in C_0^\infty(\mathbb{R}^n),$$

which is expressed the *symbol* $a(x, \xi)$. The first part of this paper is to study basic properties of symbols. We will impose the condition that $a(x, \xi)$ has polynomial growth while $|\xi| \to \infty$. When $\mathrm{Op}(a)$ is a differential operator, then $a(x, \xi)$ is indeed a polynomial, otherwise $\mathrm{Op}(a)$ is called a pseudo-differential operator. In fact, we wish to generalize the idea of differential operators to a much bigger class of operators $\mathcal{A}$ with the following properties:

1. $\mathcal{A}$ at least contains differential operators and parametrices of elliptic operators, hypoelliptic operators and convolution operators with smooth kernels.
2. $\mathcal{A}$ is closed under addition, composition, adjoint which is invariant under coordinates change.

3. $\mathcal{A}$ has the pseudo-local property, i.e.,

$$\operatorname{sing\,supp} Au \subseteq \operatorname{sing\,supp} u \qquad \text{for all} \quad A \in \mathcal{A}.$$

Here $\operatorname{sing\,supp} u$ is the *singular support of* u which is the smallest closed set C such that u is smooth off C.

4. Every element $A \in \mathcal{A}$ induces maps $A : C_0^\infty(\mathbb{R}^n) \to C^\infty(\mathbb{R}^n)$ and $A : \mathcal{E}'(\mathbb{R}^n) \to \mathcal{D}'(\mathbb{R}^n)$.

As we mentioned before, when (1.3) is elliptic, an approximate solution is given by

$$u(x) \approx \int_{\mathbb{R}^n} e^{ix\cdot\xi}\, p(x,\xi)\widehat{f}(\xi)\,\mathrm{d}\xi \tag{1.4}$$

where $p(x,\xi)$ is the inverse of $a(x,\xi)$ under Leibniz multiplication, discussed below. More precisely, to apply the argument leading to (1.4), we need the operators to satisfy

$$A_1(x, D) \circ A_2(x, D) \approx (A_1 \circ A_2)(x, D)$$

in some well-defined sense so that

$$A_1(x, D) \circ A_1^{-1}(x, D) \approx I + \text{acceptable error terms}\,.$$

In the first part of this article, we will derive an asymptotic expansion for composition of pseudo-differential operators so that these formulas become precise.

Before we go further, we observe that the operator (1.1) in $\mathbb{R}^n$ is elliptic not only with respect to $\sigma_\psi(A)$ but also with respect to other symbols, referred to as exit symbols when we interpret $\mathbb{R}^n$ as a manifold with conical exit $|x| \to \infty$, see also Remark 1.5 below.

Given a pseudo-differential operator A of order m in $\mathbb{R}^n$, where the coefficients $a_\alpha(x)$ are classical symbols in x of order zero, cf. notation in Remark 1.5 below, denote

$$\sigma_\mathrm{e}(A)(x, \xi) = \sum_{|\alpha|\leq m} a_{\alpha,(0)}(x)\xi^\alpha \qquad x \neq 0, \qquad \xi \in \mathbb{R}^n$$

and

$$\sigma_{\psi,\mathrm{e}}(A)(x, \xi) = \sum_{|\alpha|=m} a_{\alpha,(0)}(x)\xi^\alpha \qquad x \neq 0, \qquad \xi \neq 0.$$

Here $a_{\alpha,(0)}(x)$ is the homogeneous principal part of a_α in x of order 0.

An essential aspect of ellipticity with respect to the symbols $(\sigma_\psi, \sigma_e, \sigma_{\psi,e})$ is the following result.

Let $H^s(\mathbb{R}^n)$ be the standard Sobolev space on $\mathbb{R}^n$ of order $s \in \mathbb{R}$, i.e., the space of all $u \in \mathcal{S}'(\mathbb{R}^n)$ such that $\langle \xi \rangle^s (Fu)(\xi) \in L^2(\mathbb{R}^n)$. It can be shown that

$$A : H^s(\mathbb{R}^n) \to H^{s-m}(\mathbb{R}^n),$$

is continuous for all $s \in \mathbb{R}$. It is a Fredholm operator if and only if $\sigma_\psi(A)(x,\xi) \neq 0$ for $(x,\xi) \in \mathbb{R}^n \times (\mathbb{R}^n \setminus \{0\})$, $\sigma_e(A)(x,\xi) \neq 0$ for $(x,\xi) \in (\mathbb{R}^n \setminus \{0\}) \times \mathbb{R}^n$, and $\sigma_{\psi,e}(A)(x,\xi) \neq 0$ for $(x,\xi) \in (\mathbb{R}^n \setminus \{0\}) \times (\mathbb{R}^n \setminus \{0\})$.

In particular, for the operator $A = 1 - \Delta$ considered at the very beginning we have

$$\sigma_\psi(A)(\xi) = |\xi|^2, \quad \sigma_e(A) = 1 + |\xi|^2, \quad \sigma_{\psi,e}(A) = |\xi|^2. \tag{1.5}$$

Another aspect of the pseudo-differential calculus concerns boundary value problems from the point of view of expressing parametrices of boundary value problem:

$$\begin{aligned} A(x,D)u &= f &&\text{in} &&\Omega, \\ v(u) &= g &&\text{on} &&\partial\Omega \end{aligned} \tag{1.6}$$

where v represents a boundary (or trace) operator, and $\Omega \subset \mathbb{R}^n$ is a subdomain with smooth boundary.

Pseudo-differential operators in the standard sense, outlined in more details below are motivated by the task to reflect the solvability properties of elliptic differential equations and of elliptic boundary value problems. In order to prepare new ideas from the calculus of pseudo-differential operators and boundary value problems and to keep the consideration self-contained, we formulate some structures from the well-known pseudo-differential calculus, which are for scalar operators standard and may be found in papers or textbooks, cf. [33, 38, 79], but then have analogues in other contexts. For example, various modifications refer to conical exits of the configuration to infinity. In a continuation of this material this information is the background of the calculus of operators on manifolds with singularities, also motivated to solving equations, but now for configurations with conical singularities, edge or higher corner singularities. For instance, V.A. Kondratyev's paper [40] studies elliptic boundary value problems for differential operators in domains with conical singularities. The task to express parametrices gives rise to the pseudo-differential calculus for conical singularities, together with that for manifolds with smooth boundary, see L. Boutet de Monvel [9], or B.-W. Schulze [68]. Singularities of higher order appear in connection with mixed and transmission or crack problems, cf. the monographs [31, 36], or the paper [11]. It turns out that the details require new structures, not only new classes of weighted Sobolev spaces and edge spaces, as established in [66], but also new techniques of proving continuity of operators in those spaces, see the paper [75] of J. Seiler. Elements of the calculus for higher singularities are developed in [12, 14, 58, 62, 69–71, 73], and in many other papers. The corner theories, beginning with conical and edge singularities, give rise to new Mellin operators and Mellin quantizations, see G.I. Eskin's book [24] or the monographs [23, 68]. Important applications for asymptotics in many-particle systems have been started in [25]. Let us finally point out that pseudo-differential theories on manifolds with singularities seem to be at a new beginning where most of the basic problems are to be solved in future. This concerns index theories extending [9,58,59], edge conditions under non-vanishing of an analogue of the Atiyah-Bott obstruction, iterative structures with symbolic hierarchies in the sense of [70], and asymptotic properties, especially variable and boundary edge asymptotics.

1.1.2   Symbols, Operators, and Distributional Kernels

Let

$$A(x, D) = \sum_{|\alpha| \leq m} a_\alpha(x) D_x^\alpha = \sum_{\alpha_1 + \cdots + \alpha_n \leq m} a_{\alpha_1 \ldots \alpha_n}(x) \left(\frac{1}{i} \frac{\partial}{\partial x_1}\right)^{\alpha_1} \cdots \left(\frac{1}{i} \frac{\partial}{\partial x_n}\right)^{\alpha_n}$$

be a differential operator in a domain $\Omega \subseteq \mathbb{R}^n$ with coefficients $a_\alpha(x) \in C^\infty(\Omega)$, regarded as an operator

$$A(x, D) : C_0^\infty(\Omega) \to C_0^\infty(\Omega). \tag{1.7}$$

Then $A(x, D)$ can be expressed by the Fourier transform F as

$$A(x, D) = F^{-1} a(x, \xi) F \qquad \text{with} \qquad a(x, \xi) = \sum_{|\alpha| \leq m} a_\alpha(x) \xi^\alpha,$$

using the elementary identity $D_x^\alpha = F^{-1} \xi^\alpha F$. Thus

$$A(x, D)u(x) = \int_{\mathbb{R}^n} e^{ix \cdot \xi} a(x, \xi) \left\{ \int_{\mathbb{R}^n} e^{-iy \cdot \xi} u(y) dy \right\} d\!\!\!{}^-\xi$$

$$= \int_{\mathbb{R}^n} \int_{\mathbb{R}^n} e^{i(x-y) \cdot \xi} a(x, \xi) u(y) dy \, d\!\!\!{}^-\xi. \tag{1.8}$$

The expression on the right hand side is interpreted as an oscillatory integral.

Remark 1.1 Note that A through its action as a continuous operator (1.7), its symbol $a(x, \xi)$ is a unique way. In fact, using the Fourier inversion formula

$$u(x) = \int_{\mathbb{R}^n} e^{ix \cdot \xi} \widehat{u}(\xi) \, d\!\!\!{}^-\xi,$$

we obtain by applying A on both sides

$$A(x, D)u(x) = \int_{\mathbb{R}^n} \left(Ae^{ix\cdot\xi}\right)\widehat{u}(\xi)\,d\xi = \int_{\mathbb{R}^n} e^{ix\cdot\xi}\left(e^{-ix\cdot\xi}Ae^{ix\cdot\xi}\right)\widehat{u}(\xi)\,d\xi, \qquad (1.9)$$

i.e.,

$$a(x, \xi) = e^{-ix\cdot\xi}Ae^{ix\cdot\xi}. \qquad (1.10)$$

We begin by studying the appropriate classes of amplitude functions; for simplicity we also refer to them as symbols.

A function $f(\xi) \in C^\infty(\mathbb{R}^n \setminus \{0\})$ is called homogeneous of order $\mu \in \mathbb{R}$ if $f(\lambda\xi) = \lambda^\mu f(\xi)$ for all $\lambda > 0$, $\xi \in \mathbb{R}^n \setminus \{0\}$. Let $S^{(\mu)}(\Omega \times (\mathbb{R}^n \setminus \{0\}))$ be the space of all $f(x, \xi) \in C^\infty(\Omega \times (\mathbb{R}^n \setminus \{0\}))$ that are homogeneous in ξ or order μ for all $x \in \Omega$. Then

$$D_x^\alpha D_\xi^\beta S^{(\mu)}(\Omega \times (\mathbb{R}^n \setminus \{0\})) \subseteq S^{\mu-|\beta|}(\Omega \times (\mathbb{R}^n \setminus \{0\}))$$

for all $\alpha \in \mathbb{N}^m$, $\beta \in \mathbb{N}^n$.

Definition 1.2

(i) Let $\Omega \subseteq \mathbb{R}^m$ be open and $\mu \in \mathbb{R}$. Then $S^\mu(\Omega \times \mathbb{R}^n)$ denotes the space of all $a(x, \xi) \in C^\infty(\Omega \times \mathbb{R}^n)$ such that

$$\left|D_x^\alpha D_\xi^\beta a(x, \xi)\right| \leq c\left(1 + |\xi|\right)^{\mu-|\beta|} \qquad (1.11)$$

for all $\alpha \in \mathbb{N}^m$, $\beta \in \mathbb{N}^n$, $x \in K$, with arbitrary $K \Subset \Omega$, $\xi \in \mathbb{R}^n$, with constant $c = c(\alpha, \beta, K) > 0$. The number μ is called the order of the symbol a, written $\mu = \mathrm{ord}(a)$.

(ii) A symbol $a(x, \xi) \in S^\mu(\Omega \times \mathbb{R}^n)$ is called classical if there is a sequence $a_{(\mu-k)}(x, \xi) \in S^{(\mu-k)}(\Omega \times (\mathbb{R}^n \setminus \{0\}))$, $k \in \mathbb{N}$, such that for any excision function $\chi(\xi)$

$$a(x, \xi) - \sum_{k=0}^{N} \chi(\xi)a_{(\mu-k)}(x, \xi) \in S^{\mu-(N+1)}(\Omega \times \mathbb{R}^n) \qquad (1.12)$$

for all $N \in \mathbb{N}$. We denote by $S_{\mathrm{cl}}^\mu(\Omega \times \mathbb{R}^n)$ the space of all classical symbols of order μ.

The best possible constants c in (1.11) form a semi-norm system on the space $S^\mu(\Omega \times \mathbb{R}^n)$:

$$a \ \rightarrow \ \sup_{x \in K, \xi \in \mathbb{R}^n} \left| D_x^\alpha D_\xi^\beta a(x, \xi) \right| \left(1 + |\xi|\right)^{-\mu + |\beta|}.$$

It is clear that $a(x, \xi)$ belongs to $S^\mu(\Omega \times \mathbb{R}^n)$ just in case the estimates (1.11) are satisfied for a countable system of compact sets $K \subset \Omega$ and all α, β. It suffices to take all closed balls contained in Ω of rational radii and centers with rational coordinates. It is then easy to verify that $S^\mu(\Omega \times \mathbb{R}^n)$ is a Fréchet space with this semi-norm system.

The best possible constants c in (1.14) form a semi-norm system in $S^{\mu;\nu}(\mathbb{R}^m \times \mathbb{R}^n)$:

$$a \ \rightarrow \ \sup_{x, \xi \in \mathbb{R}^n} \left| D_x^\alpha D_\xi^\beta a(x, \xi) \right| \left(1 + |x|\right)^{-\nu + |\alpha|} \left(1 + |\xi|\right)^{-\mu + |\beta|}.$$

which turns $S^{\mu;\nu}(\mathbb{R}^m \times \mathbb{R}^n)$ to a Fréchet space.

Recall that a $\chi(\xi) \in C^\infty(\mathbb{R}^n)$ is called an excision function if

$$\chi(\xi) = \begin{cases} 0 & \text{if} & |\xi| < r_0 \\ 1 & \text{if} & |\xi| > r_1 \end{cases}$$

with constants $0 < r_0 < r_1 < \infty$.

Remark 1.3 The space $S_{\mathrm{cl}}^\mu(\Omega \times \mathbb{R}^n)$ is Fréchet as a projective limit with respect to the system of operators $S_{\mathrm{cl}}^\mu(\Omega \times \mathbb{R}^n) \hookrightarrow S^\mu(\Omega \times \mathbb{R}^n)$, $S_{\mathrm{cl}}^\mu(\Omega \times \mathbb{R}^n) \rightarrow S^{(\mu - j)}(\Omega \times (\mathbb{R}^n \setminus \{0\}))$, $j \in \mathbb{N}$, that determines in a unique way the component and

$$S_{\mathrm{cl}}^\mu(\Omega \times \mathbb{R}^n) \rightarrow S^{\mu - (N+1)}(\Omega \times \mathbb{R}^n), \ N \in \mathbb{N},$$

which are defined by (1.12). In this topology $S_{\mathrm{cl}}^\mu(\Omega \times \mathbb{R}^n)$ is a nuclear Fréchet space.

We write the subscript in parentheses "(cl)" at the respective spaces when a consideration is valid both in the classical and general case.

The subspaces $S_{\mathrm{cl}}^\mu(\mathbb{R}^n)$ of symbols with constant coefficients, i.e., independent of x, are closed in $S_{\mathrm{cl}}^\mu(\Omega \times \mathbb{R}^n)$. We then have

$$S_{\mathrm{cl}}^\mu(\Omega \times \mathbb{R}^n) = C^\infty(\Omega, S_{\mathrm{cl}}^\mu(\mathbb{R}^n)) = C^\infty(\Omega) \hat{\otimes}_\pi S_{\mathrm{cl}}^\mu(\mathbb{R}^n). \tag{1.13}$$

Definition 1.4

(i) $S^{\mu;\nu}(\mathbb{R}^m \times \mathbb{R}^n)$ for $\mu, \nu \in \mathbb{R}$ denotes the set of all $a(x, \xi) \in C^\infty(\mathbb{R}^m \times \mathbb{R}^n)$ such that

$$\left| D_x^\alpha D_\xi^\beta a(x, \xi) \right| \leq c\left(1 + |x|\right)^{\nu - |\alpha|} \left(1 + |\xi|\right)^{\mu - |\beta|} \tag{1.14}$$

for all $(x, \xi) \in \mathbb{R}^m \times \mathbb{R}^n$, for all $\alpha \in \mathbb{N}^m$, $\beta \in \mathbb{N}^n$ for some $c = c(\alpha, \beta) > 0$.

(ii) A symbol $a(x, \xi) \in S^{\mu;\nu}(\mathbb{R}^m \times \mathbb{R}^n)$ is called classical in x, ξ if it belongs $S^{\nu}_{\mathrm{cl}}(\mathbb{R}^m_x) \widehat{\otimes}_\pi S^{\mu}_{\mathrm{cl}}(\mathbb{R}^n_\xi) =: S^{\mu;\nu}_{\mathrm{cl}}(\mathbb{R}^m \times \mathbb{R}^n)$ where $\widehat{\otimes}_\pi$ denotes the projective tensor product between the respective Fréchet spaces of classical symbols with constant coefficients; the first factor concerns symbols in $x \in \mathbb{R}^m$, formally treated as a covariable.

Remark 1.5 The notion of homogeneous principal symbols also makes sense for $a(x, \xi) \in S^{\mu;\nu}(\mathbb{R}^m \times \mathbb{R}^n)$. First, $a(x, \xi)$ in the representation $S^{\nu}_{\mathrm{cl}}(\mathbb{R}^m_x) \widehat{\otimes}_\pi S^{\mu}_{\mathrm{cl}}(\mathbb{R}^n_\xi)$ has a homogeneous principal symbol in $\xi \neq 0$ of order μ

$$\sigma_\psi(a)(x, \xi) \in S^{\nu}_{\mathrm{cl}}(\mathbb{R}^m) \widehat{\otimes}_\pi S^{(\mu)}(\mathbb{R}^n \setminus \{0\}) \tag{1.15}$$

Moreover, $a(x, \xi)$ has a homogeneous principal symbol in $x \neq 0$ of order ν (also called principal exit symbol)

$$\sigma_{\mathrm{e}}(a)(x, \xi) \in S^{(\nu)}(\mathbb{R}^m \setminus \{0\}) \widehat{\otimes}_\pi S^{\mu}_{\mathrm{cl}}(\mathbb{R}^n), \tag{1.16}$$

and there is also a homogeneous principal symbol of (1.16) in $\xi \neq 0$, namely,

$$\sigma_{\psi,\mathrm{e}}(a)(x, \xi) \in S^{(\nu)}(\mathbb{R}^m \setminus \{0\}) \widehat{\otimes}_\pi S^{(\mu)}(\mathbb{R}^n \setminus \{0\}).$$

Note that the latter is equal to the homogeneous principal symbol of (1.15) with respect to $x \neq 0$.

This gives us altogether a triple

$$\sigma(a) = \big(\sigma_\psi(a), \sigma_{\mathrm{e}}(a), \sigma_{\psi,\mathrm{e}}(a)\big) \tag{1.17}$$

of homogeneous principal symbols. The conditions of ellipticity with respect to the components are independent.

Remark 1.6

(i) From Definitions 1.2 (i) we obtain

$$D^{\alpha}_x D^{\beta}_\xi S^{\mu}(\Omega \times \mathbb{R}^n) \subset S^{\mu - |\beta|}(\Omega \times \mathbb{R}^n)$$

for all $\alpha \in \mathbb{N}^m$, $\beta \in \mathbb{N}^n$, and (with obvious notation)

$$S^{\mu}(\Omega \times \mathbb{R}^n) \cdot S^{\nu}(\Omega \times \mathbb{R}^n) \subset S^{\mu+\nu}(\Omega \times \mathbb{R}^n)$$

for all $\mu, \nu \in \mathbb{R}$. Analogous relations hold for the classical symbols where the respective homogeneous principal symbols behave multiplicatively.

(ii) From the above-mentioned definition we see that

$$D_x^\alpha D_\xi^\beta S^{\mu;\nu}(\mathbb{R}^m \times \mathbb{R}^n) \subset S^{\mu-|\beta|;\nu-|\alpha|}(\mathbb{R}^m \times \mathbb{R}^n)$$

for all $\alpha \in \mathbb{N}^m$, $\beta \in \mathbb{N}^n$, Moreover, we have

$$S^{\mu_1;\nu_1}(\mathbb{R}^m \times \mathbb{R}^n) \cdot S^{\mu_2;\nu_2}(\mathbb{R}^m \times \mathbb{R}^n) \subset S^{\mu_1+\mu_2;\nu_1+\nu_2}(\mathbb{R}^m \times \mathbb{R}^n)$$

for all μ_j, ν_j, $j = 1, 2$. Analogous relations hold for classical symbols, where the components of the homogeneous principal symbols behave multiplicatively.

The following theorem is recommended as an exercise.

Theorem 1.7

(i) *Let* $a_j(x, \xi) \in S^{\mu_j}(\Omega \times \mathbb{R}^n)$, $j \in \mathbb{N}$, *be an arbitrary sequence, where* $\mu_j \to -\infty$ *as* $j \to \infty$. *Then there is a symbol* $a(x, \xi) \in S^\mu(\Omega \times \mathbb{R}^n)$ *with* $\mu = \max_{j \in \mathbb{N}}\{\mu_j\}$ *such that for every* M *there is an* $N(M)$ *such that for all* $N \geq N(M)$

$$a(x, \xi) - \sum_{j=0}^{N} a_j(x, \xi) \in S^{\mu-M}(\Omega \times \mathbb{R}^n).$$

The element $a(x, \xi) \in S^\mu(\Omega \times \mathbb{R}^n)$ *is uniquely determined by this property modulo* $S^{-\infty}(\Omega \times \mathbb{R}^n)$.

(ii) *Let* $a_j(x, \xi) \in S^{\mu_j;\nu_j}(\mathbb{R}^m \times \mathbb{R}^n)$, $j \in \mathbb{N}$, *for* $\mu_j \to -\infty$, $\nu_j \to -\infty$ *as* $j \to \infty$. *Then there is a symbol* $a(x, \xi) \in S^{\mu;\nu}(\mathbb{R}^m \times \mathbb{R}^n)$ *for* $\mu = \max_{j \in \mathbb{N}}\{\mu_j\}$, $\nu = \max_{j \in \mathbb{N}}\{\nu_j\}$ *and that for all* $N \geq N(M)$

$$a(x, \xi) - \sum_{j=0}^{N} a_j(x, \xi) \in S^{\mu-M;\nu-M}(\mathbb{R}^m \times \mathbb{R}^n),$$

and $a(x, \xi) \in S^{\mu;\nu}(\mathbb{R}^m \times \mathbb{R}^n)$ *is uniquely determined by this property modulo* $S^{-\infty;-\infty}(\mathbb{R}^m \times \mathbb{R}^n)$.

We call $a(x, \xi)$ an *asymptotic sum* of $a_j(x, \xi)$, $j \in \mathbb{N}$, and write

$$a \sim \sum_{j=0}^{\infty} a_j.$$

It can easily be proved that for every sequence $a_j(x, \xi) \in S^{\mu_j}(\Omega \times \mathbb{R}^n)$, $j \in \mathbb{N}$, and any choice of an excision function $\chi(\xi)$ in $\mathbb{R}^n$ there are constants $c_j > 0$, tending to ∞ sufficiently fast as $j \to \infty$, such that

$$a(x, \xi) := \sum_{j=0}^{\infty} \chi\left(\frac{\xi}{c_j}\right) a_j(x, \xi)$$

converges in $S^{\mu}(\Omega \times \mathbb{R}^n)$. For the second case, i.e., $a_j(x, \xi) \in S^{\mu_j; \nu_j}(\mathbb{R}^m \times \mathbb{R}^n)$, $j \in \mathbb{N}$, we choose an excision function $\chi(x, \xi)$ in $\mathbb{R}^{m+n}$ and find constants $c_j > 0$, tending to ∞ sufficiently fast as $j \to \infty$, such that

$$a(x, \xi) := \sum_{j=0}^{\infty} \chi\left(\frac{x, \xi}{c_j}\right) a_j(x, \xi)$$

converges in $S^{\mu; \nu}(\mathbb{R}^m \times \mathbb{R}^n)$ and represents an element with the asserted properties.

For classical symbols in Theorem 1.7 (i) we assume $a_j(x, \xi) \in S_{\mathrm{cl}}^{\mu-j}(\Omega \times \mathbb{R}^n)$, $j \in \mathbb{N}$, and in Theorem 1.7 (ii) $a_j(x, \xi) \in S_{\mathrm{cl}}^{\mu-j; \nu-j}(\mathbb{R}^m \times \mathbb{R}^n)$, $j \in \mathbb{N}$. Then the corresponding asymptotic sums are again classical.

Definition 1.8 Let $\Omega \subseteq \mathbb{R}^n$ be open and $\mu \in \mathbb{R}$. An operator of the form

$$\mathrm{Op}(a)u(x) = \iint e^{i(x-y)\cdot\xi} a(x, y, \xi) u(y) dy \,\dbar\xi \tag{1.18}$$

with a symbol (or amplitude function) $a(x, y, \xi) \in S^{\mu}(\Omega \times \Omega \times \mathbb{R}^n)$ is a pseudo-differential operator on Ω of order μ. Expression (1.18) first refers to $u \in C_0^{\infty}(\Omega)$ and is interpreted as an oscillatory integral, cf. the consideration below. For the moment we can interpret (1.18) as an iterated integral

$$\mathrm{Op}(a)u(x) = \int e^{ix\cdot\xi} \{\int e^{-iy\cdot\xi} a(x, y, \xi) u(y) dy\} \dbar\xi$$

which is convergent in the usual sense. We set

$$L^{\mu}(\Omega) = \{\mathrm{Op}(a) : a(x, y, \xi) \in S^{\mu}(\Omega \times \Omega \times \mathbb{R}^n)\}, \tag{1.19}$$

$$L_{\mathrm{cl}}^{\mu}(\Omega) = \{\mathrm{Op}(a) : a(x, y, \xi) \in S_{\mathrm{cl}}^{\mu}(\Omega \times \Omega \times \mathbb{R}^n)\}. \tag{1.20}$$

The elements of (1.20) are called classical pseudo-differential operators. Again we employ subscript "(cl)" if a consideration is valid both in the classical and the general case.

Let us set

$$L^{-\infty}(\Omega) := \bigcap_{\mu \in \mathbb{R}} L^{\mu}(\Omega).$$

For any $A \in L^{\mu}(\Omega)$ we have a continuous map

$$A : C_0^{\infty}(\Omega) \to C^{\infty}(\Omega). \tag{1.21}$$

Thus, by virtue of the Schwartz theorem, we have a distributional kernel $K_A(x, y) \in \mathcal{D}'(\Omega \times \Omega)$, where

$$\langle Au, v \rangle = \langle K_A(x, y), u(y)v(x) \rangle$$

for all $u, v \in C_0^{\infty}(\Omega)$; here $\langle \cdot, \cdot \rangle$ means the bilinear pairing. The distributional kernel K_A is in one-to-one correspondence to the respective operator A.

Definition 1.9 An operator $A \in L^{\mu}(\Omega)$ is called properly supported, if both $\operatorname{supp} K_A \cap \{M \times \Omega\}$ and $\operatorname{supp} K_A \cap \{\Omega \times M'\}$ are compact for arbitrary compact subsets $M, M' \Subset \Omega$.

Theorem 1.10 *For every $A \in L^{\mu}(\Omega)$ we have*

$$\operatorname{sing\,supp} K_A \subseteq \operatorname{diag}(\Omega \times \Omega). \tag{1.22}$$

Proof We employ the identity

$$e^{i(x-y)\cdot\xi} = |x - y|^{-2N}\left(-\Delta_{\xi}\right)^N e^{i(x-y)\cdot\xi} \tag{1.23}$$

for any $N \in \mathbb{N}$. This gives us, using integration by parts

$$\begin{aligned}
Au(x) &= \int e^{i(x-y)\cdot\xi} a(x, y, \xi)u(y)\,dy\,\text{\dj}\xi \\
&= \int |x - y|^{-2N}\left(-\Delta_{\xi}\right)^N e^{i(x-y)\cdot\xi} a(x, y, \xi)u(y)\,dy\,\text{\dj}\xi \\
&= \int e^{i(x-y)\cdot\xi}|x - y|^{-2N}\left(-\Delta_{\xi}\right)^N a(x, y, \xi)u(y)\,dy\,\text{\dj}\xi.
\end{aligned}$$

Applying Remark 1.6 (i) we have $\left(-\Delta_{\xi}\right)^N a(x, y, \xi) \in S^{\mu-2N}(\Omega \times \Omega \times \mathbb{R}^n)$ which gives us

$$K_A(x, y) = \int e^{i(x-y)\cdot\xi}|x - y|^{-2N}\left(-\Delta_{\xi}\right)^N a(x, y, \xi)u(y)\,\text{\dj}\xi.$$

This is a convergent integral as long as $x \neq y$ and N is sufficiently large, in fact, $\mu - 2N < -n$. Moreover, for every $k \in \mathbb{N}$, we have $K_A(x, y) \in C^k(\Omega \times \Omega \setminus (\mathrm{diag}(\Omega \times \Omega)))$ for any $N = N(k)$ sufficiently large. Since k is arbitrary, it follows that

$$K_A(x, y) \in C^\infty(\Omega \times \Omega \setminus (\mathrm{diag}(\Omega \times \Omega))). \tag{1.24}$$

This completes the proof of the theorem. $\qquad\square$

Corollary 1.11 *Every $A \in L^\mu(\Omega)$ can be written in the form*

$$A = A_0 + C$$

where $A_0 \in L^\mu(\Omega)$ is properly supported and $C \in L^{-\infty}(\Omega)$.

Proof In fact, looking at the expression

$$Au(x) = \iint e^{i(x-y)\cdot\xi} a(x, y, \xi) u(y) \, đ\xi$$

we can always write

$$Au(x) = \iint e^{i(x-y)\cdot\xi} \big[\omega(x, y) + (1 - \omega(x, y))\big] a(x, y, \xi) u(y) \, đ\xi$$

for arbitrary $\omega \in C_0^\infty(\Omega \times \Omega)$. Choosing ω in such a way that $\mathrm{supp}(\omega)$ is proper in the sense of Definition 1.9 and $\omega \equiv 1$ in a neighborhood of $\mathrm{diag}(\Omega \times \Omega)$ that contains $\mathrm{diag}(\Omega \times \Omega)$ in its open interior which is always possible. Then, by virtue of (1.22), $A_0 = \mathrm{Op}(\omega a)$ is properly supported while $C = \mathrm{Op}((1 - \omega)a)$ is smoothing. $\qquad\square$

Remark 1.12 Let $A \in L^\mu(\Omega)$; then (1.7) extends to a linear operator

$$A : \mathcal{E}'(\Omega) \to \mathcal{D}'(\Omega).$$

If A is properly supported then A induces continuous operator

$$A : C_0^\infty(\Omega) \to C_0^\infty(\Omega), \qquad \text{and} \qquad A : C^\infty(\Omega) \to C^\infty(\Omega)$$

which extend to linear operators

$$A : \mathcal{E}'(\Omega) \to \mathcal{E}'(\Omega), \qquad \text{and} \qquad A : \mathcal{D}'(\Omega) \to \mathcal{D}'(\Omega).$$

Example 1 A differential operator (1.3) is always properly supported. In this case,

$$\text{supp } K_A \subseteq \text{diag}(\Omega \times \Omega)$$

which entails that A is local, i.e., $\text{supp } Au \subseteq \text{supp } u$. As is well-known the differential operators are the only local operators in $L^\mu(\Omega)$.

Remark 1.13 From (1.22), we see that an $A \in L^\mu(\Omega)$ is pseudo-local, i.e.,

$$\text{sing supp } Au \subseteq \text{sing supp } u$$

for every $u \in \mathcal{E}'(\Omega)$.

Theorem 1.14 *Let $A \in L^\mu(\Omega)$ be properly supported. Then there is a unique left symbol $a(x, \xi) \in S^\mu(\Omega \times \mathbb{R}^n)$ such that $A = \text{Op}(a)$.*

Denoting by $S^\mu(\Omega \times \mathbb{R}^n)_M$ the subspace of all $a(x, \xi) \in S^\mu(\Omega \times \mathbb{R}^n)$ such that $M := \text{supp } K_{\text{Op}(a)}$, and $L^\mu(\Omega)_M$ the space of all $A \in L^\mu(\Omega)$ such that $\text{supp } K_A \subseteq M$, then the above mentioned correspondence $A \to a(x, \xi)$ gives us an isomorphism

$$L^\mu(\Omega)_M \cong S^\mu(\Omega \times \mathbb{R}^n)_M.$$

1.1.3 Double, Left and Right Symbols

For $\text{Op}(\cdot)$ to express a pseudo-differential operator in terms of a symbol, following the terminology of H. Kumano-go [42] we may admit double symbols $a(x, y, \xi) =: a_D(x, y, \xi)$ or left and right symbols $a_L(x, \xi)$ and $a_R(y, \xi)$, respectively. The following theorem tells us that $\text{Op}(a_D)$ can always be turned into $\text{Op}(a_L)$ or $\text{Op}(a_R)$ modulo smoothing operators. For the proof, we need the following result.

Lemma 1.15 *Let $a(x, y, \xi) \in S^\mu(\Omega \times \Omega \times \mathbb{R}^n)$ be a symbol with the property $|x - y|^{-2N} a(x, y, \xi) \in S^\mu(\Omega \times \Omega \times \mathbb{R}^n)$ for some $N \in \mathbb{N}$. Then there exists an $a_N(x, y, \xi) \in S^{\mu-2N}(\Omega \times \Omega \times \mathbb{R}^n)$ such that*

$$\text{Op}(a) = \text{Op}(a_N).$$

Proof Using (1.23):

$$e^{i(x-y)\cdot\xi} = |x - y|^{-2N}\left(-\Delta_\xi\right)^N e^{i(x-y)\cdot\xi}$$

and integration by parts, it follows that $\text{Op}(a) = \text{Op}(a_N)$ with

$$a_N(x, y, \xi) = \left(-\Delta_\xi\right)^N |x - y|^{-2N} a(x, y, \xi) \in S^{\mu-2N}(\Omega \times \Omega \times \mathbb{R}^n).$$

$\square$

Theorem 1.16 *To every $a(x, y, \xi) \in S^\mu(\Omega \times \Omega \times \mathbb{R}^n)$ there is an $a_L \in S^\mu(\Omega \times \mathbb{R}^n)$ with* $\mathrm{Op}(a) = \mathrm{Op}(a_L)$ *modulo $L^{-\infty}(\Omega)$ and $a_L(x, \xi)$ admits the asymptotic expansion*

$$a_L(x, \xi) \sim \sum_\alpha \frac{1}{\alpha!} D_\xi^\alpha \partial_y^\alpha a(x, y, \xi)\Big|_{y=x}. \tag{1.25}$$

In other words every $A \in L^\mu(\Omega)$ has a left symbol. Moreover, there is a right symbol $a_R(y, \xi) \in S^\mu(\Omega \times \mathbb{R}^n)$ with $\mathrm{Op}(a) = \mathrm{Op}(a_R)$ modulo $L^{-\infty}(\Omega)$ that has the asymptotic expansion

$$a_R(y, \xi) \sim \sum_\alpha \frac{1}{\alpha!} (-D_\xi)^\alpha \partial_x^\alpha a(x, y, \xi)\Big|_{x=y}. \tag{1.26}$$

Proof Applying the Taylor expansion of $a(x, y, \xi)$ on $\mathrm{diag}(\Omega \times \Omega)$ we obtain

$$a(x, y, \xi) = \sum_{|\alpha| \leq M} \frac{1}{\alpha!} (y - x)^\alpha \partial_y^\alpha a(x, y, \xi)\Big|_{y=x} + R_M(x, y, \xi) \tag{1.27}$$

with a remainder term $R_M(x, y, \xi) \in S^\mu(\Omega \times \Omega \times \mathbb{R}^n)$. For every N we can choose M so large such that

$$|x - y|^{-2N} R_M(x, y, \xi) \in S^\mu(\Omega \times \Omega \times \mathbb{R}^n).$$

Applying Op on both sides of (1.27) we obtain from Lemma 1.15, together with the identity $\mathrm{Op}\big((y - x)^\alpha a_R\big) = \mathrm{Op}\big(D_\xi^\alpha a_R\big)$ for any symbol a_R,

$$\mathrm{Op}(a) = \sum_{|\alpha| \leq M} \frac{1}{\alpha!} \mathrm{Op}\Big(D_\xi^\alpha \partial_y^\alpha a(x, y, \xi)\Big|_{y=x}\Big) + \mathrm{Op}(a_N)$$

with some $a_N \in S^{\mu-2N}(\Omega \times \Omega \times \mathbb{R}^n)$. If we form (1.25) by carrying out the asymptotic sum we obtain immediately

$$\mathrm{Op}(a) - \mathrm{Op}(a_L) = \mathrm{Op}(\widetilde{a}_N)$$

for another $\widetilde{a}_N \in S^{\mu-2N}(\Omega \times \Omega \times \mathbb{R}^n)$. This is true for every N and hence $\mathrm{Op}(a) - \mathrm{Op}(a_L) \in L^{-\infty}(\Omega)$. The second statement can be proved in an analogous manner by interchanging the role of x and y. The proof of the theorem is therefore complete. $\square$

1.1.4 Formal Adjoints, Compositions

Theorem 1.17 *Let $A \in L^\mu(\Omega)$ and let A^* be its formal adjoint, defined by*

$$\left(Au, v\right) = \left(u, A^*v\right) \qquad \text{for all} \quad u, v \in C_0^\infty(\Omega)$$

with the L^2 scalar product $(\cdot, \cdot)$. Then $A^* \in L^\mu(\Omega)$. If $A = \mathrm{Op}(a)$ modulo $L^{-\infty}(\Omega)$, for some $a(x, \xi) \in S^\mu(\Omega \times \mathbb{R}^n)$, we have $A^* = \mathrm{Op}(a^*)$ modulo $L^{-\infty}(\Omega)$ for an $a^*(x, \xi) \in S^\mu(\Omega \times \mathbb{R}^n)$ such that

$$a^*(x, \xi) \sim \sum_\alpha \frac{1}{\alpha!} D_\xi^\alpha \partial_x^\alpha \overline{a(x, \xi)}. \tag{1.28}$$

Proof The behavior of operators in $L^{-\infty}(\Omega)$ under formal adjoints is obvious. Thus we may assume $A = \mathrm{Op}(a)$ with a left symbol $a(x, \xi)$. Then

$$A^*v(x) = \iint e^{i(x-y)\cdot\xi} \overline{a(y, \xi)} v(y) dy \, \dbar\xi$$

shows $A^* \in L^\mu(\Omega)$ including (1.28) as a consequence of Theorem 1.16. The conclusion of the theorem follows immediately. $\square$

Now we come to the most important result in this section.

Theorem 1.18 *Let $A_1 \in L^{\mu_1}(\Omega)$ and $A_2 \in L^{\mu_2}(\Omega)$ and A_1 or A_2 be properly supported. Then $A_1 A_2 \in L^{\mu_1+\mu_2}(\Omega)$. If $A_1 = \mathrm{Op}(a_1)$ modulo $L^{-\infty}(\Omega)$, $a_1(x, \xi) \in S^{\mu_1}(\Omega \times \mathbb{R}^n)$, and $A_2 = \mathrm{Op}(a_2)$ modulo $L^{-\infty}(\Omega)$, $a_2(x, \xi) \in S^{\mu_2}(\Omega \times \mathbb{R}^n)$, then $A_1 A_2 = \mathrm{Op}(a)$ modulo $L^{-\infty}(\Omega)$ for a symbol $a(x, \xi) \in S^{\mu_1+\mu_2}(\Omega \times \mathbb{R}^n)$,*

$$a(x, \xi) \sim \sum_\alpha \frac{1}{\alpha!} \left(D_\xi^\alpha a_1(x, \xi)\right)\left(\partial_x^\alpha a_2(x, \xi)\right). \tag{1.29}$$

Proof Let us assume, for instance, that A_1 is properly supported. Write A_2 in the form $A_2 = \mathrm{Op}(\widetilde{a}_2)$ modulo $L^{-\infty}(\Omega)$ for a symbol $\widetilde{a}_2(y, \xi) \in S^{\mu_2}(\Omega \times \mathbb{R}^n)$. Then, according to (1.26),

$$\widetilde{a}_2(y, \xi) \sim \sum_\alpha \frac{1}{\alpha!} \left(-D_\xi\right)^\alpha \left(\partial_y^\alpha a_2(y, \xi)\right). \tag{1.30}$$

The relations

$$\mathrm{Op}(\widetilde{a}_2)u(y) = F_{\xi\to y}^{-1} F_{y\to\xi} \left\{\widetilde{a}_2(y, \xi)u(y)\right\}$$

and

$$\mathrm{Op}(a_1)v(x) \;=\; F^{-1}_{\xi \to x} a_1(x,\xi)\{F_{y \to \xi} v(y)\}$$

yield

$$\mathrm{Op}(a_1)\mathrm{Op}(\widetilde{a}_2)u(x) \;=\; \iint e^{i(x-y)\cdot \xi} a_1(x,\xi)\widetilde{a}_2(y,\xi)u(y)\,dy\,\dj\xi.$$

In order to obtain an asymptotic formula for $a(x,\xi)$ it suffices to apply (1.25) to $a_1(x,\xi)\widetilde{a}_2(y,\xi)$, which gives us

$$a(x,\xi) \;\sim\; \sum_{\alpha} \frac{1}{\alpha!}\partial_\xi^\alpha \{a_1(x,\xi)\,D_x^\alpha \widetilde{a}_2(x,\xi)\}$$

$$\sim\; \sum_{\alpha,\beta} \frac{1}{\alpha!\beta!}\partial_\xi^\alpha \{a_1(x,\xi)\,(-\partial_\xi)^\beta D_x^{\alpha+\beta} a_2(x,\xi)\}.$$

This was used in the second relation (1.30). Now we employ the Leibniz rule

$$\partial_\xi^\alpha\big(f(\xi)g(\xi)\big) \;=\; \sum_{\gamma+\delta=\alpha} \frac{\alpha!}{\gamma!\delta!}\{\partial_\xi^\gamma f(\xi)\}\{\partial_\xi^\delta g(\xi)\}$$

and the binomial formula

$$(x+y)^k \;=\; \sum_{\beta+\delta=\mathbf{k}} \frac{\mathbf{k}!}{\beta!\delta!} x^\delta y^\beta$$

for arbitrary $x,y \in \mathbb{R}^n$, $\mathbf{k} \in \mathbb{N}^n$. Then, in particular, for $x = -y = e = (1,1,\dots,1)$ we get

$$(e-e)^k \;=\; \sum_{\beta+\delta=\mathbf{k}} \frac{\mathbf{k}!}{\beta!\delta!} e^\delta (-e)^\beta \;=\; \mathbf{k}! \sum_{\beta+\delta=\mathbf{k}} \frac{(-1)^{|\beta|}}{\beta!\delta!} \;=\; \delta_{0,\mathbf{k}}$$

with $\delta_{0,\mathbf{k}} = 1$ for $\mathbf{k} = (0,0,\dots,0)$, $\delta_{0,\mathbf{k}} = 0$ for $\mathbf{k} \neq 0$. Thus

$$a(x,\xi) \sim \sum_{\alpha,\beta,\gamma,\delta,\gamma+\delta=\alpha} \frac{1}{\beta!\gamma!\delta!} \{\partial_\xi^\gamma a_1(x,\xi)\}\{(-\partial_\xi)^\beta \partial_\xi^\delta D_x^{\alpha+\beta} a_2(x,\xi)\}$$

$$= \sum_{\beta,\gamma,\delta} \frac{(-1)^{|\beta|}}{\beta!\gamma!\delta!} \{\partial_\xi^\gamma a_1(x,\xi)\}\{\partial_\xi^{\beta+\delta} D_x^{\beta+\gamma+\delta} a_2(x,\xi)\}$$

$$= \sum_\gamma \frac{1}{\gamma!} \sum_k \left\{ \sum_{\beta+\delta=k} \frac{(-1)^{|\beta|}}{\beta!\delta!} \right\} [\partial_\xi^\gamma a_1(x,\xi)] \times [\partial_\xi^k D_x^{k+\gamma} a_2(x,\xi)]$$

This can be simplified to (1.29) and the proof of the theorem is therefore complete. $\square$

Remark 1.19 Observe that

$$a_1(x,\xi)a_2(x,\xi) = a_1(x,\xi)\#a_2(x,\xi) \mod S^{\mu_1+\mu_2-1}(\Omega \times \mathbb{R}^n).$$

In other words, the commutator $A_1 A_2 - A_2 A_1$ belongs to $L^{\mu_1+\mu_2-1}(\Omega \times \mathbb{R}^n)$.

The asymptotic sum in (1.29) will also be called the *Leibniz product* between a_1 and a_2, written

$$a(x,\xi) = a_1(x,\xi)\#a_2(x,\xi). \tag{1.31}$$

$a(x,\xi)$ is unique modulo $S^{-\infty}(\Omega \times \mathbb{R}^n)$.

1.1.5 Recovering of Symbols

Let us set $e_\xi(x) = e^{ix\cdot\xi}$ and consider a properly supported operator $A \in L^\mu(\Omega)$. Then

$$\widetilde{a}(x,\xi) := e_{-\xi}(x)Ae_\xi \tag{1.32}$$

is a smooth function in (x,ξ).

Theorem 1.20 *Let $A \in L^\mu(\Omega)$ be properly supported. Then (1.32) is a double symbol of A satisfying $A = \mathrm{Op}(\widetilde{a})$. If the operator is given as $A = \mathrm{Op}(a)$ for an $a(x,y,\xi) \in S^\mu(\Omega \times \Omega \times \mathbb{R}^n)$ then (1.32) has the asymptotic expansion*

$$\widetilde{a}(x,\xi) \sim \sum_\alpha \frac{1}{\alpha!} D_\xi^\alpha \partial_y^\alpha a(x,y,\xi)\Big|_{y=x}.$$

Proof Let us apply A to $u \in C_0^\infty(\Omega)$ and use the Fourier inversion formula to write

$$u(x) = \int_{\mathbb{R}^n} e_\xi(x)\widehat{u}(\xi)\,\text{\dj}\xi.$$

Then

$$Au(x) = \int_{\mathbb{R}^n} \big(Ae_\xi\big)(x)\widehat{u}(\xi)\,\text{\dj}\xi,$$

because the integrals converge in $C^\infty(\Omega)$. Thus

$$Au(x) = \int e^{ix\cdot\xi}\big\{e_{-\xi}(x)(Ae_\xi)(x)\big\}\widehat{u}(\xi)\,\text{\dj}\xi$$

$$= \int e^{ix\cdot\xi}\widetilde{a}(x,\xi)\Big\{\int e^{-iy\cdot\xi}u(y)dy\Big\}\text{\dj}\xi.$$

We will show $\widetilde{a}(x,\xi) \in S^\mu(\Omega \times \mathbb{R}^n)$. Then $\widetilde{a}(x,\xi)$ is a left symbol of A. Since A is properly supported, we may start with $A = \text{Op}(a)$ for an $a(x,y,\xi) \in S^\mu(\Omega \times \Omega \times \mathbb{R}^n)$ which has proper support in (x,y). We have

$$\widetilde{a}(x,\xi) := \iint e^{-ix\cdot\xi}e^{i(x-y)\cdot\eta}a(x,y,\eta)e^{iy\cdot\xi}dy\,\text{\dj}\eta \tag{1.33}$$

as an iterated integral, where the integration in y is taken over a compact set when $x \in K$ for any $K \Subset \Omega$. The expression (1.33) can be regarded as an oscillatory integral dependent on the parameter $x \in K$. Substituting $z = x - y$, $\zeta = \eta - \xi$ we obtain

$$\widetilde{a}(x,\xi) := \iint a(x,x+z,\xi+\zeta)e^{-iz\cdot\zeta}dz\,\text{\dj}\zeta.$$

Taylor expansion of $a(x,x+z,\xi+\zeta)$ in ζ at $\zeta = 0$ gives us

$$a(x,x+z,\xi+\zeta) = \sum_{|\alpha|\leq N-1}\frac{1}{\alpha!}\big(\partial_\xi^\alpha a\big)(x,x+z,\xi)\zeta^\alpha + R_N(x,x+z,\xi,\zeta)$$

with

$$R_N(x,x+z,\xi,\zeta) = \sum_{|\alpha|=N}\frac{N\zeta^\alpha}{\alpha!}\int_0^1(1-t)^{N-1}\big(\partial_\xi^\alpha a\big)(x,x+z,\xi+t\zeta)dt.$$

From the Fourier inversion formula it follows that

$$\iint \big(\partial_\xi^\alpha a\big)(x,x+z,\xi)\zeta^\alpha e^{-iz\cdot\zeta}\,dz\,\text{\dj}\zeta = \partial_\xi^\alpha D_z^\alpha a(x,x+z,\xi)\Big|_{z=0}.$$

Thus

$$\tilde{a}(x,\xi) := b_N(x,\xi) + \iint R_N(x, x+z, \xi, \zeta)e^{-iz\cdot\zeta}\,dz\,đ\zeta \tag{1.34}$$

for

$$b_N(x,\xi) = \sum_{|\alpha|\leq N-1} \frac{1}{\alpha!}\partial_\xi^\alpha D_y^\alpha a(x,y,\xi)\Big|_{y=x} \in S^\mu(\Omega \times \mathbb{R}^n).$$

If we show that the second summand on the right hand side of (1.34) is a symbol $C_N(x,\xi)$ of order μ_N tending to $-\infty$ as $N \to \infty$, then we obtain $\tilde{a}(x,\xi) \in S^\mu(\Omega \times \mathbb{R}^n)$ and at the same time the asserted asymptotic expansion. We have

$$C_N(x,\xi) = \sum_{|\alpha|=N} \frac{N}{\alpha!}\int_0^1 (1-t)^{N-1}\Big\{\iint e^{-iz\cdot\zeta}\zeta^\alpha\partial_\xi^\alpha a(x, x+z, \xi+t\zeta)\,dz\,đ\zeta\Big\}dt.$$

Let us consider

$$r_{\alpha,t}(x,\xi) = \iint e^{-iz\cdot\zeta}\zeta^\alpha\partial_\xi^\alpha a(x, x+z, \xi+t\zeta)\,dz\,đ\zeta.$$

Integrating by parts yields

$$r_{\alpha,t}(x,\xi) = \iint e^{-iz\cdot\zeta}\partial_\xi^\alpha D_z^\alpha a(x, x+z, \xi+t\zeta)\,dz\,đ\zeta.$$

Choosing an $m \in \mathbb{N}$ and applying again integration by parts, using

$$\langle\zeta\rangle^{-2m}\langle D_z\rangle^{2m}e^{iz\cdot\zeta} = e^{-iz\cdot\zeta}$$

it follows that $r_{\alpha,t}(x,\xi)$ equals a finite sum of expressions of the form

$$r_{\alpha,\beta,t}(x,\xi) = \iint e^{-iz\cdot\zeta}\langle\zeta\rangle^{-2m}\partial_\xi^\alpha D_z^{\alpha+\beta}a(x, x+z, \xi+t\zeta)\,dz\,đ\zeta,$$

where $|\beta| \leq 2m$. Let us write $r_{\alpha,\beta,t} = p_{\alpha,\beta,t}+q_{\alpha,\beta,t}$, where $p_{\alpha,\beta,t}$ is defined as the integral over $\{(z,\zeta) : |\zeta| \leq |\xi|/2\}$ and $q_{\alpha,\beta,t}$ as the integral over the complement. If $|\zeta| \leq |\xi|/2$ then

$$\frac{|\xi|}{2} \leq |\xi + t\zeta| \leq \frac{3|\xi|}{2}.$$

Since the integration in $p_{\alpha,\beta,t}$ with respect to ζ is taken over a domain of measure $\leq c|\xi|^n$ for a constant c, we obtain

$$\left| p_{\alpha,\beta,t}(x,\xi) \right| \leq c\langle\zeta\rangle^{\mu-N+n} \tag{1.35}$$

for another constant c which is independent of ξ and t. For $|\zeta| > |\xi|/2$ we have

$$\left| \partial_\xi^\alpha D_z^{\alpha+\beta} a(x, x+z, \xi+t\zeta) \right| \leq c\langle\zeta\rangle^{\mu-N}$$

for $\mu - N \geq 0$ and $\leq$ constant for $\mu - N < 0$. Thus, if $h := \max\{\mu - N, 0\}$, we get

$$\left| q_{\alpha,\beta,t}(x,\xi) \right| \leq c \int_{|\zeta| > |\xi|/2} \langle\zeta\rangle^{-2m} \langle\zeta\rangle^\ell d\zeta.$$

For m so large that $\ell - 2m + n + 1 < 0$ it follows that

$$\left| q_{\alpha,\beta,t}(x,\xi) \right| \leq c\langle\xi\rangle^{\ell-2m+n+1} \int \langle\zeta\rangle^{-n-1} d\zeta \leq c\langle\xi\rangle^{\ell-2m+n+1} \tag{1.36}$$

with different constants c that are independent of ξ and t, $0 \leq t \leq 1$. Since m is arbitrary, (1.35) and (1.36) imply the required estimate for the remainder, with $\mu_N = \mu - N + n \to -\infty$ and $N \to \infty$. The proof of the theorem is therefore complete. $\quad\square$

Theorem 1.21 *For $\Omega \subseteq \mathbb{R}^n$ and $\mu \in \mathbb{R}$, the map*

$$\mathrm{Op} : S^\mu(\Omega \times \mathbb{R}^n) \to L^\mu(\Omega) \tag{1.37}$$

induces an (algebraic) isomorphism

$$S^\mu(\Omega \times \mathbb{R}^n)/S^{-\infty}(\Omega \times \mathbb{R}^n) \cong L^\mu(\Omega)/L^{-\infty}(\Omega).$$

Proof The map (1.37) induces an linear operator

$$\mathrm{Op} : S^{-\infty}(\Omega \times \mathbb{R}^n) \to L^{-\infty}(\Omega). \tag{1.38}$$

Therefore, (1.37) together with (1.38) induces a map between the respective quotient spaces

$$S^\mu(\Omega \times \mathbb{R}^n)/S^{-\infty}(\Omega \times \mathbb{R}^n) \to L^\mu(\Omega)/L^{-\infty}(\Omega). \tag{1.39}$$

This map is surjective, since $L^\mu(\Omega)$ can be equivalently characterized as

$$L^{\mu}(\Omega) = \{\mathrm{Op}(a) + C : a(x,\xi) \in S^{\mu}(\Omega \times \mathbb{R}^{n}), \ C \in L^{-\infty}(\Omega)\},$$

cf. relation (1.19), i.e., it suffices to take left symbols, in order to reach $L^{\mu}(\Omega)/L^{-\infty}(\Omega)$.

On the other hand, (1.39) is surjective, because of the fact, that $A = \mathrm{Op}(a)$ can be written as $A = A_0 + C$ for a properly supported $A_0 \in L^{\mu}(\Omega)$ and $C \in L^{-\infty}(\Omega)$.

Writing with notation from Theorem 1.20

$$\begin{aligned} L^{\mu}(\Omega) &= L^{\mu}(\Omega)_M + L^{-\infty}(\Omega), \\ S^{\mu}(\Omega \times \mathbb{R}^{n}) &= S^{\mu}(\Omega \times \mathbb{R}^{n})_M + S^{-\infty}(\Omega \times \mathbb{R}^{n}), \end{aligned} \tag{1.40}$$

where $M \Subset \Omega$ is an arbitrary compact. We see at the same time that the map

$$L^{\mu}(\Omega)_M \ \to \ S^{\mu}(\Omega \times \mathbb{R}^{n})_M$$

induces a map

$$L^{-\infty}(\Omega)_M \ \to \ S^{-\infty}(\Omega \times \mathbb{R}^{n})_M.$$

Thus, since

$$L^{\mu}(\Omega)/L^{-\infty}(\Omega) \ = \ L^{\mu}(\Omega)_M/L^{-\infty}(\Omega)_M,$$

and

$$S^{\mu}(\Omega \times \mathbb{R}^{n})/S^{-\infty}(\Omega \times \mathbb{R}^{n}) \ = \ S^{\mu}(\Omega \times \mathbb{R}^{n})_M/S^{-\infty}(\Omega \times \mathbb{R}^{n})_M,$$

the map (1.39) is an isomorphism. $\qquad\square$

1.1.6 Symbol Classes $S^{\mu}_{\rho,\delta}(\Omega \times \mathbb{R}^{n})$

Definition 1.22 Let $\Omega \subseteq \mathbb{R}^{n}$ be open and $\mu \in \mathbb{R}$. Let $S^{\mu}_{\rho,\delta}(\Omega \times \mathbb{R}^{n})$ be the collection of all $p(x,\xi)$ in $C^{\infty}(\Omega \times \mathbb{R}^{n})$ such that

$$|D_x^{\alpha} D_{\xi}^{\beta} p(x,\xi)| \le c\langle\xi\rangle^{\mu-\rho|\beta|+\delta|\alpha|}. \tag{1.41}$$

where $\langle\xi\rangle = (1 + |\xi|^2)^{\frac{1}{2}}$, $\forall \alpha, \beta \in (\mathbb{Z}_+)^{n}$, $\mu, \rho, \delta \in \mathbb{R}$, and suppose $0 \le \rho, \delta \le 1$, $x \in K$, with arbitrary $K \Subset \Omega$, $\xi \in \mathbb{R}^{n}$, with constant $c = c(\alpha, \beta, K) > 0$. μ is called the order of the symbol p.

Proposition 1.23 *From the preceding definition, one obtains*

$$S^{\mu}_{\rho_1,\delta_1}(\Omega \times \mathbb{R}^n)S^{\nu}_{\rho_2,\delta_2}(\Omega \times \mathbb{R}^n) \subset S^{\mu+\nu}_{\rho,\delta}(\Omega \times \mathbb{R}^n)$$

for all μ, $\nu \in \mathbb{R}$, where $\rho = \min\{\rho_1, \rho_2\}$, $\delta = \max\{\delta_1, \delta_2\}$.

Proof Let $f \in S^{\mu}_{\rho_1,\delta_1}$, $g \in S^{\nu}_{\rho_2,\delta_2}$ and $k, k', \ell, \ell' \in (\mathbb{Z}_+)^n$ are multi-indices, then by the Leibniz rule, we have

$$
\begin{aligned}
|D^{\alpha}_x D^{\beta}_{\xi}(fg)| &= |D^{\alpha}_x(\sum_{\{k+k'=\beta\}} \frac{\beta!}{k!k'!} D^k_{\xi} f \cdot D^{k'}_{\xi} g)| \\[2mm]
&= |\sum_{\{k+k'=\beta\}} \frac{\beta!}{k!k'!} D^{\alpha}_x(D^k_{\xi} f \cdot D^{k'}_{\xi} g)| \\[2mm]
&= |\sum_{\{k+k'=\beta\}} \frac{\beta!}{k!k'!} \sum_{\{\ell+\ell'=\alpha\}} \frac{\alpha!}{\ell!\ell'!} D^{\ell}_x D^k_{\xi} f \cdot D^{\ell'}_x D^{k'}_{\xi} g| \\[2mm]
&= |\sum_{\{k+k'=\beta\}} \sum_{\{\ell+\ell'=\alpha\}} \frac{\beta!}{k!k'!} \frac{\alpha!}{\ell!\ell'!} D^{\ell}_x D^k_{\xi} f \cdot D^{\alpha-\ell}_x D^{k'}_{\xi} g| \\[2mm]
&\leq \sum_{\{k+k'=\beta\}} \sum_{\{\ell+\ell'=\alpha\}} \frac{\beta!}{k!k'!} \frac{\alpha!}{\ell!\ell'!} |D^{\ell}_x D^k_{\xi} f||D^{\ell'}_x D^{k'}_{\xi} g| \\[2mm]
&\leq \sum_{\{k+k'=\beta\}} \sum_{\{\ell+\ell'=\alpha\}} \frac{\beta!}{k!k'!} \frac{\alpha!}{\ell!\ell'!} (c_{\mu}\langle\xi\rangle^{\mu-\rho_1|k|+\delta_1|\ell|})(c_{\nu}\langle\xi\rangle^{\nu-\rho_2|k'|+\delta_2|\ell'|}) \\[2mm]
&\leq \sum_{\{k+k'=\beta\}} \sum_{\{\ell+\ell'=\alpha\}} \frac{\beta!}{k!k'!} \frac{\alpha!}{\ell!\ell'!} (c_{\mu}\langle\xi\rangle^{\mu-\rho|k|+\delta|\ell|})(c_{\nu}\langle\xi\rangle^{\nu-\rho|k'|+\delta|\ell'|}) \\[2mm]
&= \sum_{\{k+k'=\beta\}} \sum_{\{\ell+\ell'=\alpha\}} \frac{\beta!}{k!k'!} \frac{\alpha!}{\ell!\ell'!} (c_{\mu}c_{\nu})\langle\xi\rangle^{\mu+\nu-\rho(|k|+|k'|)+\delta(|\ell|+|\ell'|)} \\[2mm]
&= \sum_{\{k+k'=\beta\}} \sum_{\{\ell+\ell'=\alpha\}} \frac{\beta!}{k!k'!} \frac{\alpha!}{\ell!\ell'!} (c_{\mu}c_{\nu})\langle\xi\rangle^{\mu+\nu-\rho|\beta|+\delta|\alpha|} \\[2mm]
&= C\langle\xi\rangle^{\mu+\nu-\rho|\beta|+\delta|\alpha|},
\end{aligned}
$$

where $C = \sum_{\{k+k'=\beta\}} \sum_{\{\ell+\ell'=\alpha\}} \frac{\beta!}{k!k'!} \frac{\alpha!}{\ell!\ell'!} c_{\mu}c_{\nu}$. By definition, we have

$$S^{\mu}_{\rho_1,\delta_1}(\Omega \times \mathbb{R}^n)S^{\nu}_{\rho_2,\delta_2}(\Omega \times \mathbb{R}^n) \subset S^{\mu+\nu}_{\rho,\delta}(\Omega \times \mathbb{R}^n).$$

$\square$

Before we go further, let us introduce one more symbol class. Let $\mu \in \mathbb{R}$ and $0 \le \rho, \delta_1, \delta_2 \le 1$. We say that a function $p(x, y, \xi)$ in $C^{\infty}(\Omega \times \Omega \times \mathbb{R}^n)$ lies in the *symbol class* $S^{\mu}_{\rho,\delta_1,\delta_2}(\Omega \times \Omega \times \mathbb{R}^n)$ if there is a compact set $K \subset \Omega$ such that the support of $p(x, y, \xi)$ is contained in $K \times K \times \mathbb{R}^n$, and for any multi-indices α, β, γ there is a constant $c = c(\alpha, \beta, \gamma)$ such that

$$|D_x^\alpha D_y^\gamma D_\xi^\beta p(x, y, \xi)| \le c\langle\xi\rangle^{\mu-\rho|\beta|+\delta_1|\alpha|+\delta_2|\gamma|}.$$

Proposition 1.24 *If $p(x, y, \xi) \in S^{\mu}_{\rho,\delta_1,\delta_2}$ with $0 \le \delta_2 < \rho \le 1$, then the operator $Pu = \iint e^{i(x-y)\cdot\xi} p(x, y, \xi)u(y)\, dy\, d̶\xi$ belong to $OPS^{\mu}_{\rho,\delta}$, with $\delta = \max(\delta_1, \delta_2)$. In fact $P = q(x, D)$ where $q(x, \xi)$ has the asymptotic expansion*

$$q(x, \xi) \sim \sum_{\alpha \ge 0} \frac{i^{|\alpha|}}{\alpha!} D_\xi^\alpha D_y^\alpha p(x, y, \xi)\Bigg|_{y=x}. \tag{1.42}$$

Proposition 1.25 *If $p(x, D) \in OPS^{\mu}_{\rho,\delta}$, $0 \le \delta < \rho \le 1$, then*

$$p(x, D)^* = p^*(x, D) \in OPS^{\mu}_{\rho,\delta} \tag{1.43}$$

with

$$p^*(x, \xi) \sim \sum_{\alpha \ge 0} \frac{i^{|\alpha|}}{\alpha!} D_\xi^\alpha D_x^\alpha p(x, \xi)^*. \tag{1.44}$$

Theorem 1.26 *Let $\mu_j \in \mathbb{R}$, $0 \le \rho_j, \delta_j \le 1$ and $P_j \in OPS^{\mu_j}_{\rho_j,\delta_j}$, for $j = 1, 2$. Then $P_1 P_2 \in OPS^{\mu_1+\mu_2}_{\rho,\delta}$ with $\delta = \max(\delta_1, \delta_2)$, $\rho = \min(\rho_1, \rho_2)$, $0 \le \delta_2 < \rho \le 1$. If $P_j = Op(p_j)$, $p_j(x, \xi) \in S^{\mu_j}_{\rho_j,\delta_j}(\Omega \times \mathbb{R}^n)$ for $j = 1, 2$, then*

$$P_1 P_2 = Op(q)$$

for a symbol $q(x, \xi) \in S^{\mu_1+\mu_2}_{\rho,\delta}(\Omega \times \mathbb{R}^n)$ and

$$q(x, \xi) \sim \sum_{\alpha \ge 0} \frac{1}{\alpha!} D_\xi^\alpha p_1(x, \xi)\partial_x^\alpha p_2(x, \xi).$$

Proof Let $P = P_1 P_2$. Note that $Op(p_2) = Op(p_2)^{**} = Op(p_2^*)^*$. So

$$P = P_1 P_2 = Op(p_1)Op(p_2) = Op(p_1)Op(p_2^*)^*$$

for symbols $p_1(x, \xi) \in S^{\mu_1}_{\rho_1, \delta_1}(\Omega \times \mathbb{R}^n)$ and $\overline{p_2^*(y, \xi)} \in S^{\mu_2}_{\rho_2, \delta_2}(\Omega \times \mathbb{R}^n)$. By Proposition 1.23, $p(x, y, \xi) = p_1(x, \xi)\overline{p_2^*(y, \xi)} \in S^{\mu_1 + \mu_2}_{\rho, \delta_1, \delta_2}(\Omega \times \Omega \times \mathbb{R}^n)$, where $\rho = \min\{\rho_1, \rho_2\}$. Analogous to the argument as in the proof of Proposition 1.23,

$$Pu = \iint e^{i(x-y)\cdot\xi} p_1(x, \xi)\overline{p_2^*(y, \xi)}u(y)dy\,\text{d}\xi.$$

By Proposition 1.24, there is a symbol $q(x, \xi)$ in $S^{\mu}_{\rho, \delta}(\Omega \times \mathbb{R}^n)$, where $\delta = \max(\delta_1, \delta_2)$, with $P = Op(q)$, So P is in $OPS^{\mu_1 + \mu_2}_{\rho, \delta}$. By (1.42) and (1.44),

$$q(x, \xi) \sim \sum_{\alpha \geq 0} \frac{1}{\alpha!} D^{\alpha}_{\xi} \partial^{\alpha}_y p(x, y, \xi)\Big|_{y=x} = \sum_{\alpha \geq 0} \frac{1}{\alpha!} D^{\alpha}_{\xi} \partial^{\alpha}_y \left[p_1(x, \xi)\overline{p_2^*(y, \xi)} \right]\Big|_{y=x}$$

$$\sim \sum_{\alpha \geq 0} \frac{1}{\alpha!} D^{\alpha}_{\xi} \partial^{\alpha}_y \left[p_1(x, \xi) \left(\sum_{\beta \geq 0} \frac{1}{\beta!} D^{\beta}_{\xi} \partial^{\beta}_y p_2(y, \xi) \right) \right]\Big|_{y=x}$$

$$= \sum_{\alpha, \beta \geq 0} \frac{1}{\alpha!\beta!} D^{\alpha}_{\xi} \left[p_1(x, \xi) \left(D^{\beta}_{\xi} \partial^{\alpha+\beta}_x p_2(x, \xi) \right) \right].$$

Applying the Leibniz rule on $D^{\alpha}_{\xi} \left[p_1(x, \xi) \left(D^{\beta}_{\xi} \partial^{\alpha+\beta}_x p_2(x, \xi) \right) \right]$, it follows that

$$q(x, \xi) \sim \sum_{\alpha, \beta \geq 0} \frac{1}{\alpha!\beta!} \sum_{\substack{\gamma+\delta=\alpha \\ \gamma, \delta \in (\mathbb{Z}_+)^n}} \frac{\alpha!}{\gamma!\delta!}(D^{\gamma}_{\xi} p_1(x, \xi)) \left(D^{\delta}_{\xi} \left(D^{\beta}_{\xi} \partial^{\alpha+\beta}_x p_2(x, \xi) \right) \right)$$

$$= \sum_{\beta, \delta, \gamma \geq 0} \frac{1}{\gamma!\delta!\beta!}(D^{\gamma}_{\xi} p_1(x, \xi)) \left(D^{\beta+\delta}_{\xi} \partial^{\gamma+\delta+\beta}_x p_2(x, \xi) \right)$$

$$= \sum_{\gamma \geq 0} \frac{1}{\gamma!} \sum_{k} \{ \sum_{\beta+\delta=k} \frac{1}{\delta!\beta!} \}(D^{\gamma}_{\xi} p_1(x, \xi)) \left(D^{k}_{\xi} \partial^{\gamma+k}_x p_2(x, \xi) \right)$$

$$= \sum_{\gamma \geq 0} \frac{1}{\gamma!} \sum_{k} \{ \sum_{\beta+\delta=k} (-1)^{|\beta|} \sum_{\beta+\delta=k} \frac{(-1)^{|\beta|}}{\delta!\beta!} \}(D^{\gamma}_{\xi} p_1(x, \xi)) \left(D^{k}_{\xi} \partial^{\gamma+k}_x p_2(x, \xi) \right)$$

$$= \sum_{\gamma \geq 0} \frac{1}{\gamma!}(D^{\gamma}_{\xi} p_1(x, \xi))\partial^{\gamma}_x p_2(x, \xi).$$

$\square$

We need to prove the following main theorem of this section.

Theorem 1.27 *Let* $p(x, \xi) \in S^{\mu}_{\rho, \delta}(\mathbb{R}^n \times \mathbb{R}^n)$ *be a symbol where* $0 \leq \delta < \rho \leq 1$ *and* $P = Op(p)$, *then*

$$P : H^s(\mathbb{R}^n) \to H^{s-\mu}(\mathbb{R}^n)$$

for every $s, \mu \in \mathbb{R}$.

Proof Let the function

$$b(\zeta, \xi) = \int e^{-ix\cdot\zeta} p(x, \xi)\, dx$$

satisfy

$$|b(\zeta, \xi)| \le c_N (1 + |\zeta|)^{-N}(1 + |\xi|)^\mu \tag{1.45}$$

for every $N \in \mathbb{N}$ with some constant $c_N > 0$, then we have $\zeta^\alpha b(\zeta, \xi) = \int e^{-2\pi i x\zeta} D_x^\alpha p(x, \xi)\, dx$ for every $\alpha \in \mathbb{N}^n$ which gives us for arbitrary $M \in \mathbb{N}$

$$|(1 + |\zeta|^2)^M b(\zeta, \xi)| = \left| \int e^{-ix\cdot\zeta} (1 + (\tfrac{1}{i})^2 \Delta_x)^M p(x, \xi)\, dx \right|$$

$$= \left| \int e^{-ix\cdot\zeta} \sum_{k=0}^{M} \binom{M}{k} ((\tfrac{1}{i})^2 \Delta_x)^k p(x, \xi)\, dx \right|$$

$$= \left| \int e^{-ix\cdot\zeta} \sum_{k=0}^{M} \binom{M}{k} \sum_{|\beta|=k} \frac{k!}{\beta!} D_x^{2\beta} p(x, \xi)\, dx \right|$$

where $\beta \in \mathbb{N}^n$

$$= \left| \int e^{-ix\cdot\zeta} \sum_{|\beta|=0}^{M} \frac{M!}{(M - |\beta|!)\beta!} D_x^{2\beta} p(x, \xi)\, dx \right|$$

$$\le \int |e^{-ix\cdot\zeta} \sum_{|\beta|=0}^{M} \frac{M!}{(M - |\beta|!)\beta!} D_x^{2\beta} p(x, \xi)|\, dx \tag{1.46}$$

$$\le \int \sum_{|\beta|=0}^{M} \frac{M!}{(M - |\beta|!)\beta!} |D_x^{2\beta} p(x, \xi)|\, dx$$

$$\le \int \sum_{|\beta|=0}^{M} \frac{M!}{(M - |\beta|!)\beta!} c'(1 + |\xi|)^{\mu + \delta|2\beta|}\, dx$$

$$\le \int \sum_{|\beta|=0}^{M} \frac{M!}{(M - |\beta|!)\beta!} c'(1 + |\xi|)^{\mu + 1\cdot|2M|}\, dx$$

$$\le c_M (1 + |\xi|)^{\mu + 2M}$$

where $c_M = \int \sum_{|\beta|=0}^{M} \frac{M!}{(M-|\beta|!)\beta!} c' dx > 0$. Since $(1+|\zeta|)^N (1+|\zeta|^2)^{-M} < \tilde{c}$ for $M \geq \frac{N}{2}$, then we have

$$|b(\zeta,\xi)| \leq c_M (1+|\zeta|^2)^{-M} (1+|\xi|)^{\mu+2M}$$

$$\leq c_M \tilde{c} (1+|\zeta|)^{-N} (1+|\xi|)^{\mu+2M}$$

$$= c_N (1+|\zeta|)^{-N} (1+|\xi|)^{\mu+2M},$$

we get (1.45).

Next we set $K(x,\eta) = b(\eta-\xi,\xi)(1+|\xi|)^{-(s+2M)}(1+|\eta|)^{s-\mu}$. Then (1.46) together with

$$\left(\frac{1+|\eta|}{1+|\xi|}\right)^s \leq (1+|\xi-\eta|)^{|s|}$$

gives us the estimate

$$|K(\xi,\eta)| = |b(\eta-\xi,\xi)(1+|\xi|)^{-(s+2M)}(1+|\eta|)^{s-\mu}|$$

$$\leq c_N (1+|\eta-\xi|)^{-N}(1+|\xi|)^{\mu+2M}(1+|\xi|)^{-(s+2M)}(1+|\eta|)^{s-\mu}$$

$$\leq c_N (1+|\xi-\eta|)^{-N} \left(\frac{(1+|\eta|)}{(1+|\xi|)}\right)^{s-\mu}$$

$$\leq c_N (1+|\xi-\eta|)^{-N}(1+|\xi-\eta|)^{|s-\mu|}$$

$$= c_N (1+|\xi-\eta|)^{-N+|s-\mu|}$$

when we choose $N > n + |s-\mu|$, we obtain

$$\int |K(\xi,\eta)|d\xi \leq c, \ \int |K(\xi,\eta)|d\eta \leq c \tag{1.47}$$

for all η and ξ, respectively, for some constant $c > 0$.

Applying now the Fourier transform $\mathcal{F}_{x\to\eta}$ to $Pu(x)$, $u \in C_0^\infty(\mathbb{R}^n)$, we obtain

$$(\widehat{Pu})(\eta) = \iint e^{ix\cdot(\xi-\eta)} p(x,\xi)\widehat{u}(\xi)d\xi dx = \int b(\eta-\xi,\xi)\widehat{u}(\xi)d\xi.$$

this yields for any $v \in H^{\mu-s}(\mathbb{R}^n)$

$$\int \widehat{Pu}(\eta)\widehat{v}(\eta)d\eta = \iint b(\eta-\xi,\xi)\widehat{v}(\eta)\widehat{u}(\xi)d\xi d\eta$$

$$= \iint K(\xi,\eta)\widehat{v}(\eta)(1+|\eta|)^{\mu-s}\widehat{u}(\xi)(1+|\xi|)^{s+2M}d\xi d\eta.$$

We then obtain

$$
\begin{aligned}
|\langle Pu, v\rangle| &= |\langle \widehat{Pu}, \widehat{v}\rangle| \\
&= \left| \int \widehat{Pu}(\eta)\widehat{v}(\eta)\,d\eta \right| \\
&= \left| \iint K(\xi, \eta)\widehat{v}(\eta)(1 + |\eta|)^{\mu-s}\widehat{u}(\xi)(1 + |\xi|)^{s+2M}\,d\xi\,d\eta \right| \\
&\leq \iint |K(\xi, \eta)||\widehat{v}(\eta)||(1 + |\eta|)^{\mu-s}||\widehat{u}(\xi)||(1 + |\xi|)^{s+2M}|\,d\xi\,d\eta \\
&\leq \left\{ \iint |K(\xi, \eta)|(1 + |\eta|)^{2(\mu-s)}|\widehat{v}(\eta)|^2\,d\xi\,d\eta \right\}^{\frac{1}{2}} \\
&\quad \cdot \left\{ \iint |K(\xi, \eta)||\widehat{u}(\xi)|^2(1 + |\xi|)^{2(s+2M)}\,d\xi\,d\eta \right\}^{\frac{1}{2}} \\
&\leq \left\{ c\,\|v\|_{\mu-s}^2 \right\}^{\frac{1}{2}} \cdot \left\{ c\,\|u\|_{s+2M}^2 \right\}^{\frac{1}{2}} \\
&\leq c\,\|v\|_{\mu-s}\,\|u\|_{s+2M} \\
&= c\,\|v\|_{\mu-s}\,\|u\|_{\widetilde{s}},
\end{aligned}
$$

where $\widetilde{s} = s + 2M \in \mathbb{R}$, and we have used (4.15). Then it follows that

$$
\sup_{\substack{v \in H^{\mu-s}(\mathbb{R}^n) \\ v \neq 0}} \frac{|\langle Pu, v\rangle|}{\|v\|_{\mu-s}} \leq c\|u\|_{\widetilde{s}}.
$$

By Theorem 1.27 , we obtain $\|Pu\|_{s-\mu} \leq c\|u\|_{\widetilde{s}}$, for some $c > 0$. The proof of this theorem is therefore complete. $\qquad\square$

1.2 Operators on Manifolds

Pseudo-differential operators are known to be meaningful also on smooth manifolds. First observe that a diffeomorphism $\chi : \Omega \to \widetilde{\Omega}$ between open sets $\Omega, \widetilde{\Omega}$ in $\mathbb{R}^n$ induces an operator push forward defined by $\chi_* A = (\chi^*)^{-1}A\chi^*$. In the later notation we assume $A \in L^\mu(\Omega)$, $A : C_0^\infty(\Omega) \to C^\infty(\Omega)$, and $\chi^* : C_0^\infty(\widetilde{\Omega}) \to C_0^\infty(\Omega)$, $C^\infty(\widetilde{\Omega}) \to C^\infty(\Omega)$ are the corresponding function pull-backs.

Let $\varphi(x, x', \xi)$ be a phase function on $\Omega \times \Omega \times (\mathbb{R}^n \setminus \{0\})$, which means a real-valued $\varphi(x, x', \xi) \in C^\infty(\Omega \times \Omega \times (\mathbb{R}^n \setminus \{0\}))$ such that $\varphi(x, x', \delta\xi) = \delta\varphi(x, x', \xi)$ for all $\delta \in \mathbb{R}_+$, $(x, x', \xi) \in \Omega \times \Omega \times (\mathbb{R}^n \setminus \{0\})$, and $d_{x,x',\xi}\varphi \neq 0$ for all $(x, x', \xi) \in \Omega \times \Omega \times (\mathbb{R}^n \setminus \{0\})$, where $d_{x,x',\xi}$ means the gradient with respect to corresponding variables.

An analogous definition makes sense when we replace $\Omega \times \Omega \times (\mathbb{R}^n \setminus \{0\}) \ni (x, x', \xi)$ by $\Omega \times (\mathbb{R}^n \setminus \{0\}) \ni (x, \xi)$ for an open $\Omega \subseteq \mathbb{R}^m$.

An example is the function

$$\varphi(x, x', \xi) := (x - x') \cdot \xi$$

occurring in (1.18). In this case we have some extra properties that make φ to an operator phase function in the sense that we require the conditions $d_{x,\xi}\varphi(x, x', \xi) \neq 0$ and $d_{x',\xi}\varphi(x, x', \xi) \neq 0$ for all $(x, x', \xi) \in \Omega \times \Omega \times (\mathbb{R}^n \setminus \{0\})$. Let $\varphi(x, x', \xi)$ be an operator phase function and set

$$C_\varphi := \{(x, x', \xi) \in \Omega \times \Omega \times (\mathbb{R}^n \setminus \{0\}) : d_\xi \varphi(x, x', \xi) = 0\}, \tag{1.48}$$

$d_\xi \varphi := (\frac{\partial \varphi}{\partial \xi_1}, \ldots, \frac{\partial \varphi}{\partial \xi_n})$, and denote by $\pi : C_\varphi \to \Omega \times \Omega$ the projection $(x, x', \xi) \to (x, x')$. In order to associate with an expression

$$Au := \iint e^{i\varphi(x,x',\xi)} a(x, x', \xi) u(x') dx' \, d\xi \tag{1.49}$$

an operator $A : C_0^\infty(\Omega) \to C^\infty(\Omega)$ we interpret (1.49) which as an oscillatory integral.

In order to explain the procedure we consider for the moment a phase function $\varphi(x, \xi)$ over $\Omega \times (\mathbb{R}^n \setminus \{0\})$, i.e., replace the former (x, x') by x.

Lemma 1.28 *For any phase function* $\varphi(x, \xi)$ *on* $\Omega \times (\mathbb{R}^n \setminus \{0\})$, $\Omega \subset \mathbb{R}^m$ *open, there exists an operator*

$$L = \sum_{j=1}^m a_j(x, \xi) \frac{\partial}{\partial x_j} + \sum_{k=1}^n b_k(x, \xi) \frac{\partial}{\partial \xi_k} + c(x, \xi)$$

with coefficients

$$a_j \in S^{-1}(\Omega \times \mathbb{R}^n), \ b_k(x, \xi) \in S^0(\Omega \times \mathbb{R}^n), \ c(x, \xi) \in S^{-1}(\Omega \times \mathbb{R}^n)$$

such that the formal transposed operator $^{\mathrm{t}}L$

$$^{\mathrm{t}}Lv = -\sum_{j=1}^m \frac{\partial}{\partial x_j}(a_j v) - \sum_{k=1}^n \frac{\partial}{\partial \xi_k}(b_k v) + cv$$

satisfies

$$^{\mathrm{t}}Le^{i\varphi} = e^{i\varphi}.$$

Proof It suffices to find $M := {}^{\mathrm{t}}L$ such that $Me^{i\varphi} = e^{i\varphi}$, and we set

$$M := -i\psi(1-\omega)\Big\{ \sum_{j=1}^{m} \frac{\partial\varphi}{\partial x_j}\frac{\partial}{\partial x_j} + \sum_{k=1}^{n} |\xi|^2 \frac{\partial\varphi}{\partial \xi_k}\frac{\partial}{\partial \xi_k} \Big\} + \omega.$$

Here

$$\psi(x,\xi) = \Big\{ \sum_{j=1}^{m} \Big|\frac{\partial\varphi}{\partial x_j}\Big|^2 + \sum_{k=1}^{n} |\xi|^2 \Big|\frac{\partial\varphi}{\partial \xi_k}\Big|^2 \Big\}^{-1}$$

which belongs to $C^\infty(\Omega \times (\mathbb{R}^n \setminus \{0\}))$. Then we choose an $\omega(\xi) \in C_0^\infty(\mathbb{R}^n)$ such that $\omega(\xi) = 1$ in a neighbourhood of $\xi = 0$. From the homogeneity of φ in ξ of order 1 we then easily obtain the asserted properties. $\qquad\square$

Set

$$I_\varphi(au) := \iint e^{i\varphi(x,\xi)} a(x,\xi) u(x)\, dx\, \text{d}\xi \tag{1.50}$$

for $a(x,\xi) \in S^\mu(\Omega \times \mathbb{R}^n)$. The integral converges for $\mu < -n$. For arbitrary μ the method of regularization of (1.50) as an oscillatory integral consists of replacing (1.50) by

$$\iint e^{i\varphi(x,\xi)} L^N (a(x,\xi) u(x))\, dx\, \text{d}\xi$$

for $N \in \mathbb{N}$ so large that $\mu - N < -n$, where we use that $L^N S^\mu(\Omega \times \mathbb{R}^n) \subset S^{\mu-N}(\Omega \times \mathbb{R}^n)$. The result is independent of the choice of L or N, as we see from the following proposition.

Proposition 1.29 *Let $a(x,\xi) \in S^\mu(\Omega \times \mathbb{R}^n)$ and $a_\varepsilon(x,\xi) := \omega(\varepsilon\xi)a(x,\xi)$ for some $\omega \in C_0^\infty(\mathbb{R}^n)$, such that $\omega \equiv 1$ in a neighbourhood of $\xi = 0$, $\varepsilon > 0$, we have*

$$\iint e^{i\varphi(x,\xi)} L^N (a(x,\xi) u(x))\, dx\, \text{d}\xi = \lim_{\varepsilon\to 0} I_\varphi(a_\varepsilon u)$$

for every N such that $\mu - N < -n$.

Proof Because of $a_\varepsilon(x,\xi) \in S^{-\infty}(\Omega \times \mathbb{R}^n)$ the integral $I_\varphi(a_\varepsilon u)$ converges in the usual sense. Moreover, integration by parts gives us

$$I_\varphi(a_\varepsilon u) := \iint e^{i\varphi(x,\xi)} L^N (\omega(\varepsilon\xi)a(x,\xi) u(x))\, dx\, \text{d}\xi. \tag{1.51}$$

Because of $|D_\xi^\alpha \omega(\varepsilon\xi)| \le c_\alpha \langle\xi\rangle^{-|\alpha|}$ for all α for constants c_α independent of ε for $0 < \varepsilon \le 1$, by Lebesgue's theorem on dominated convergence we may pass to the limit for $\varepsilon \to 0$ under the integral on the right-hand side of (1.51). This completes the proof. $\qquad\square$

Theorem 1.30 *Let $\varphi(x, x', \xi)$ be a phase function which is linear in ξ, and assume $\pi C_\varphi = \mathrm{diag}\,(\Omega \times \Omega)$. Then for any $a(x, x', \xi) \in S_{(cl)}^\mu (\Omega \times \Omega \times \mathbb{R}^n)$ the operator*

$$Au := \iint e^{i\varphi(x,x',\xi)} a(x, x', \xi) u(x') dx' \bar{d}\xi \tag{1.52}$$

belongs to $L_{(cl)}^\mu (\Omega)$.

Proof We will show that there is a neighbourhood V of $\mathrm{diag}(\Omega \times \Omega)$ and a C^∞ map $g : V \to \mathrm{GL}(n, \mathbb{R})$ such that

$$\varphi(x, x', g(x, x')\xi) = (x - x') \cdot \xi \quad \text{for all } (x, x') \in V. \tag{1.53}$$

Here $g(x, x) = {}^t(d_{x,\xi}\varphi(x, x', \xi)|_{x'=x})^{-1}$, where $d_{x,\xi}\varphi$ is the matrix of derivatives $\partial_{x_j}\partial_{\xi_k}\varphi$. This implies

$$\det g(x, x') \det d_{x,\xi}\varphi(x, x', \xi)|_{x'=x} = 1. \tag{1.54}$$

Because of linearity of φ in ξ we have

$$\varphi(x, x', \xi) = \sum_{j=1}^n \varphi_j(x, x')\xi_j$$

where $\varphi_j(x, x) = 0$ for $j = 1, \dots, n$. Moreover, $\varphi_j(x, x') = 0$ for all $j = 1, \dots, n$ implies $x = x'$. In addition let us verify that

$$\det \left(\frac{\partial \varphi_j(x, x')}{\partial x_k}\right)\Big|_{x'=x} \neq 0. \tag{1.55}$$

In fact we first observe that differentiating relation $\varphi(x, x, \xi) = 0$ with respect to x yields $d_x\varphi|_{x'=x} = -d_{x'}\varphi|_{x'=x}$. By assumption, we have $d_{x,x',\xi}\varphi = 0$ for $\xi \neq 0$. Then $d_\xi\varphi(x, x', \xi) = 0$ implies $d_x\varphi(x, x', \xi) \neq 0$ for $\xi \neq 0$. Thus, for every $\xi \neq 0$ there is a k such that $\sum_{j=1}^n \frac{\partial \varphi_j}{\partial x_k}\xi_j|_{x'=x} \neq 0$. This implies (1.55). Applying Hadamard's Lemma for $\Phi(x, x') := (\varphi_1(x, x'), \dots, \varphi_n(x, x'))$ we find a neighbourhood $\tilde{V}$ of $\mathrm{diag}\,(\Omega \times \Omega)$ such that $\Phi(x, x') = G(x, x')(x - x')$ for a C^∞ matrix function $G(x, x') = (G_{jk}(x, x'))_{j,k=1,\dots,n}$ on $\tilde{V}$, where

$$G_{jk}(x, x) = \frac{\partial \varphi_j(x, x')}{\partial x_k}\Big|_{x'=x}. \tag{1.56}$$

Relations (1.55), (1.56) imply that for some neighbourhood V of $\mathrm{diag}(\Omega \times \Omega)$ the matrix $G(x, x')$ is non-singular for all $(x, x') \in V$. Setting $g(x, x') := {}^t G^{-1}(x, x')$ we get (1.53) from

$$\varphi(x, x', g(x, x')\xi) = \langle \Phi(x, x'), {}^t G^{-1}(x, x')\xi \rangle =$$
$$\langle G(x, x')(x - x'), {}^t G^{-1}(x, x')\xi \rangle = (x - x') \cdot \xi. \tag{1.57}$$

Then (1.54) is a consequence of (1.56).

The distributional kernel $K_A(x, x')$ of the operator (1.52) has the property (1.22), cf. (1.48). Thus, for any $\omega(x, x') \in C^\infty(\Omega \times \Omega)$ properly supported in V and $\omega \equiv 1$ close to $\mathrm{diag}(\Omega \times \Omega)$, the operator

$$Cu(x) := \iint e^{i\varphi(x,x',\xi)}(1 - \omega(x, x'))a(x, x', \xi)u(x')dx'\,\text{đ}\xi$$

belongs to $L^{-\infty}(\Omega)$. Thus, it remains to characterize

$$A_0 u(x) := \iint e^{i\varphi(x,x',\xi)}\omega(x, x')a(x, x', \xi)u(x')dx'\,\text{đ}\xi.$$

Using (1.53) and substituting $\eta = g(x, x')\xi$ gives us

$$A_0 u(x) := \iint e^{i(x-x')\eta}\omega(x, x')a(x, x', g(x, x')\eta)|\det g(x, x')|u(x')dx'\,\text{đ}\eta.$$

Finally we obtain $A_0 \in L^\mu_{(\mathrm{cl})}(\Omega)$ because of $a(x, x', g(x, x')\eta) \in S^\mu_{(\mathrm{cl})}(\Omega \times \Omega \times \mathbb{R}^n_\eta)$. $\square$

Theorem 1.31 *For a diffeomorphism $\chi : \Omega \to \tilde{\Omega}$ the operator push forward χ_* induces isomorphisms*

$$\chi_* : L^\mu_{(\mathrm{cl})}(\Omega) \to L^\mu_{(\mathrm{cl})}(\Omega)$$

for every $\mu \in \mathbb{R}$. If $A = \mathrm{Op}(a) \bmod L^{-\infty}(\Omega)$, $a = a(x, \xi) \in S^\mu(\Omega \times \mathbb{R}^n)$, and writing $\tilde{A} := \chi_* A$ as $\tilde{A} = \mathrm{Op}(\tilde{a}) \bmod L^{-\infty}(\tilde{\Omega})$, $\tilde{a} = \tilde{a}(\tilde{x}, \tilde{\xi}) \in S^\mu(\tilde{\Omega} \times \mathbb{R}^n)$, then we have

$$\tilde{a}(\tilde{x}, \tilde{\xi})_{\tilde{x}=\chi(x)} \sim \sum_{\alpha \in \mathbb{N}^n} \frac{1}{\alpha!}(\partial^\alpha_\xi a)(x, {}^t d\chi(x)\tilde{\xi})\Phi(x, \tilde{\xi}) \tag{1.58}$$

for

$$\Phi(x, \tilde{\xi}) = D_z^\alpha e^{i\delta(x,z)\tilde{\xi}}|_{z=x}, \quad \alpha \in \mathbb{N}^n, \tag{1.59}$$

which is a polynomial of degree $\leq |\alpha|/2$, *for* $\delta(x, z) := \chi(z) - \chi(x) - d\chi(x)(z - x)$ *with* $d\chi(x)$ *being the Jacobian of* χ *at* $x \in \Omega$.

Proof Since χ_* induces an isomorphism $L^{-\infty}(\Omega) \to L^{-\infty}(\tilde{\Omega})$, it suffices to consider $A \in L^\mu_{(\mathrm{cl})}(\Omega)$ in the form

$$Au(x) = \iint e^{i(x-x')\cdot\xi} a(x, \xi)u(x')dx'd\!\!\!/\xi.$$

For notational convenience throughout the proof we set $y := \tilde{x}$, $y' := \tilde{x}'$, $\eta := \tilde{\xi}$. Then writing $u := \chi^* v$, $x = \chi^{-1}(y)$, $x' = \chi^{-1}(y')$ from A we obtain

$$Bv(y) = \iint e^{i\varphi(y,y',\xi)} a(\chi^{-1}(y), \xi)|\det d\chi^{-1}(y')|v(y')dy'd\!\!\!/\xi$$

where the phase function $\varphi(y, y', \xi) = (\chi^{-1}(y) - \chi^{-1}(y'))\xi$ satisfies the assumptions of Theorem 1.30. It follows that $B \in L^\mu(\tilde{\Omega})$. For the matrix function $g(y, y')$ in the proof of Theorem 1.30 the phase function takes the form $\varphi(y, y', \xi) = (y - y')g^{-1}(y, y')\xi$. Setting $\eta := g^{-1}(y, y')\xi$, $h(y, y') := |\det g(y, y')||\det d\chi^{-1}(y')|$ it follows that

$$Bv(y) = \iint e^{i(y-y')\cdot\eta} a(\chi^{-1}(y), g(y, y')\eta)h(y, y')v(y')dy'd\!\!\!/\eta.$$

Theorem 1.16 yields an asymptotic expansion for a left symbol $b(y, \eta)$ of B, namely,

$$b(y, \eta) \sim \sum_{\alpha \in \mathbb{N}^n} \frac{1}{\alpha!} \partial_\eta^\alpha D_{y'}^\alpha a(\chi^{-1}(y), g(y, y')\eta)h(y, y')|_{y'=y}. \tag{1.60}$$

In other words we obtain $B = \mathrm{Op}(b) \bmod L^{-\infty}(\tilde{\Omega})$. The summands belonging to α have the form

$$c(y, y')\eta^\gamma (\partial_\xi^\beta a)(\chi^{-1}(y), g(y, y')\eta)|_{y'=y},$$

where $c(y, y')$ only depends on χ, not on the symbol a, and the multi-indices γ, β satisfy the relations $|\beta| \leq 2|\alpha|$, $|\gamma| + |\alpha| \leq |\beta|$. The first estimate is obvious. The second one follows from the fact that $\partial_{y'}$ in the α–th term of (1.60) does not change the difference $|\beta| - |\gamma|$ while the derivatives ∂_η raise $|\beta| - |\gamma|$ by 1. Formula (1.54) means in our case $g(y, y) = ({}^t d\chi^{-1}(y))^{-1}$. Let us now rearrange (1.60) by taking together all summands with the same β. We then obtain an asymptotic sum of the form

$$b(y, \eta) \sim \sum_{\beta \in \mathbb{N}^n} \frac{1}{\beta!} (\partial_\xi^\beta a)(\chi^{-1}(y), ({}^t d\chi^{-1}(y))^{-1}\eta)\Psi_\beta(y, \eta)$$

for polynomials $\Psi_\beta(y, \eta)$ in η of degree $\leq |\beta|/2$, not depending on the symbol of A where $\Psi_0 = 1$. Inserting $y = \chi(x)$ yields

$$b(y, \eta)|_{y=\chi(x)} = \sum_{\beta \in \mathbb{N}^n} \frac{1}{\beta!} (\partial_\xi^\beta a)(x, {}^t d\chi(x)\eta)\Phi_\beta(x, \eta)$$

where $\Phi_\beta(x, \eta)$ is a polynomial in η of degree $\leq |\beta|/2$, with coefficients in $C^\infty(\Omega)$. The polynomials are independent of the specific operator A. Therefore, they can be calculated by inserting differential operators A. In such a case we have, cf. formula (1.10),

$$b(y, \eta)|_{y=\chi(x)} = e^{-iy\cdot\eta} B e^{iy\cdot\eta}|_{y=\chi(x)} = e^{-i\chi(x)\eta} a(z, D_z)e^{i\chi(z)\eta}|_{z=x}. \tag{1.61}$$

We now write $\chi(z) = \chi(x) + d\chi(x)(z - x) + \delta(x, z)$ and play in (1.60) the identity

$$\chi(z)\eta = \chi(x)\eta + z\,{}^t d\chi(x)\eta + \delta(x, z)\eta - x\,{}^t d\chi(x)\eta.$$

Thus

$$b(y, \eta)|_{y=\chi(x)} = e^{-ix\,{}^t d\chi(x)\eta} a(z, D_z)\{e^{iz\,{}^t d\chi(x)\eta} e^{i\delta(x,z)\eta}\}|_{z=x}.$$

Applying the Leibniz rule for a differential operator $a(x, D_x) = \sum_{|\alpha|\leq\mu} a_\alpha(x)D_x^\alpha$ for $a(x, \xi) = \sum_{|\alpha|\leq\mu} a_\alpha(x)\xi^\alpha$, namely,

$$a(x, D_x)(uv)(x) = \sum_\alpha \{(\partial_\xi^\alpha a)(x, D_x)u(x)\}D_x^\alpha v(x)$$

and using relation $a(x, D_x)e^{ix\cdot\xi} = e^{ix\cdot\xi}a(x, \xi)$ we obtain formula (1.59) for the function Φ. $\qquad\square$

Let us now draw some conclusions of Theorem 1.31. First, since (1.59) is a polynomial in $\tilde{\xi}$ of degree $\leq |\alpha|/2$ and $(\partial_\xi^\alpha a)(x, {}^t d\chi(x)\tilde{\xi})$ is a symbol of order $\mu - |\alpha|$, the order of the summands in (1.58) tend to $-\infty$ as $|\alpha| \to \infty$, cf. Remark 1.6 (i). Thus, by virtue of Theorem 1.7 the asymptotic sum (1.58) exists.

For $\alpha = 0$, we have $\Phi_0 = 1$, and hence we have

$$\tilde{a}(\tilde{x}, \tilde{\xi}) = a(x, \xi) \bmod S^{\mu-1}(\Omega \times \mathbb{R}^n)$$

for

$$(\tilde{x}, \tilde{\xi}) = (\chi(x), {}^{t}d\chi^{-1}(x)\xi). \tag{1.62}$$

For classical symbols $a(x, \xi)$ it follows that $\tilde{a}_{(\mu)}(\tilde{x}, \tilde{\xi}) = a_{(\mu)}(x, \xi)$ when $(\tilde{x}, \tilde{\xi})$ is related to (x, ξ) via (4.63).

Remark 1.32 Since (4.63) just corresponds to the transition map $T^*\Omega \to T^*\tilde{\Omega}$ between the cotangent bundles of Ω and $\tilde{\Omega}$, respectively, the homogeneous principal symbols transform as functions on the corresponding cotangent bundles minus zero section, later on indicated by $\setminus 0$.

In order to fix notation we also call the transformation $a(x, \xi) \to \tilde{a}(\tilde{x}, \tilde{\xi})$ in (1.58) the symbol push forward and write similarly as for operators

$$\chi_* : S^{\mu}_{(\mathrm{cl})}(\Omega \times \mathbb{R}^n) \to S^{\mu}_{(\mathrm{cl})}(\tilde{\Omega} \times \mathbb{R}^n)$$

which is well-defined modulo smoothing remainders. Let us now turn to operators on a (paracompact) smooth manifold M with a fixed Riemannian metric. We fix a locally finite open covering $(U_j)_{j \in \mathbb{N}}$ of M by coordinate neighbourhoods, a subordinate partition of unity $(\varphi_j)_{j \in \mathbb{N}}$ and a system of function $(\varphi'_j)_{j \in \mathbb{N}}$, $\varphi'_j \in C_0^{\infty}(U_j)$, such that $\varphi_j \prec \varphi'_j$ for all j; here $f \prec g$ for functions f, g means that $g \equiv 1$ on supp f. Given a chart $\chi : U \to \Omega$ on M we define

$$L^{\mu}_{(\mathrm{cl})}(U) := (\chi^{-1})_* L^{\mu}_{(\mathrm{cl})}(\Omega)$$

where the push forward has an analogous meaning as above for a diffeomorphism between open sets in $\mathbb{R}^n$. Moreover, if $\tilde{\chi} : U \to \tilde{\Omega}$ is another chart then we have

$$L^{\mu}_{(\mathrm{cl})}(U) = (\tilde{\chi}^{-1})_* L^{\mu}_{(\mathrm{cl})}(\tilde{\Omega}).$$

In other words the spaces $L^{\mu}_{(\mathrm{cl})}(U)$ are invariantly defined. We now define

$$L^{\mu}_{(\mathrm{cl})}(M) := \Big\{ \sum_{j \in \mathbb{N}} \varphi_j A_j \varphi'_j + C : A_j \in L^{\mu}_{(\mathrm{cl})}(U_j), C \in L^{-\infty}(M) \Big\}$$

where $L^{-\infty}(M)$ is the space of all smoothing operators on M. Those are defined as operators of the form $Cu(x) = \int_M c(x, x')u(x')dx'$ where dx' is associated with the Riemannian metric.

1.3   **Twisted Symbolic Estimates**

The *standard* concept of pseudo-differential operators admits various generalizations, motivated by corresponding applications. By *standard* we understand operators with symbols as in Definition 1.2 (i), also denoted by $S^{\mu}_{1,0}(\Omega \times \mathbb{R}^n)$. There are also the well-known classes $S^{\mu}_{\rho,\delta}(\Omega \times \mathbb{R}^n)$, e.g., for $0 < \delta \le \rho \le 1$, where the symbolic estimates (1.11) are generalized to replacing $(1 + |\xi|)^{\mu-|\beta|}$ by $(1 + |\xi|)^{\mu+\delta|\alpha|-\rho|\beta|}$. There is another generalization which appears in boundary value problems and also in operators on a manifold with singularities, say, edges, where the values of the symbols belong to $\mathcal{L}(H, \widetilde{H})$ for some Hilbert spaces H, $\widetilde{H}$ (or Fréchet spaces) which are endowed with groups of isomorphisms. We need a few aspects of this concept for boundary value problems below, and later on also for operators on manifolds with edge and corner singularity.

An elementary approach is to iterate pseudo-differential operators with symbols

$$p(x, y, \xi, \eta) \in S^{\mu}(\mathbb{R}^{n+q}_{x,y} \times \mathbb{R}^{n+q}_{\xi,\eta})$$

by applying first Op_x, i.e., consider

$$\mathrm{Op}_x(p)(y, \eta) =: a(y, \eta)$$

which is a $(y, \eta)-$dependent family of operators in $L^{\mu}(\mathbb{R}^n)$, i.e., operator-valued, and then pass to $\mathrm{Op}_y(a)$. Such an iteration makes sense, in particular, in boundary and transmission problems.

Assume for convenience that the symbol p is independent of x. Then

$$\mathrm{Op}_x(p)(y, \eta) : H^s(\mathbb{R}^n) \to H^{s-\mu}(\mathbb{R}^n)$$

is a family of continuous operators for every $s \in \mathbb{R}$.

Let us consider the group of isomorphisms

$$\kappa_\lambda : H^s(\mathbb{R}^n) \to H^s(\mathbb{R}^n), \qquad \lambda \in \mathbb{R}_+,$$

defined by

$$(\kappa_\lambda u)(x) := \lambda^{\frac{n}{2}} u(\lambda x).$$

Lemma 1.33 *We have*

$$\left\| \kappa_{\langle\eta\rangle}^{-1} \mathrm{Op}_x(p)(y,\eta)\kappa_{\langle\eta\rangle} \right\|_{\mathcal{L}(H^s(\mathbb{R}^n), H^{s-\mu}(\mathbb{R}^n))} \leq c\langle\eta\rangle^\mu$$

for all $(y,\eta) \in M \times \mathbb{R}^n$, $M \Subset \mathbb{R}^n$, *and some constant* $c = c(M) > 0$. *Here* $\kappa_{\langle\eta\rangle}^{-1} = \left(\kappa_{\langle\eta\rangle}\right)^{-1}$.

Proof First observe that for any $f(\xi) \in S^\mu(\mathbb{R}_\xi^n)$ and $s \in \mathbb{R}$ we have

$$\left\| \mathrm{Op}_x(f)u \right\|_{H^{s-\mu}(\mathbb{R}^n)}^2 = \int \langle\xi\rangle^{2(s-\mu)} |f(\xi)\widehat{u}(\xi)|^2 \, \mathrm{d}\xi$$

$$\leq \sup_{\xi \in \mathbb{R}^n} \langle\xi\rangle^{-2\mu} |f(\xi)|^2 \|u\|_{H^s(\mathbb{R}^n)}^2,$$

i.e.,

$$\left\| \mathrm{Op}_x(f)u \right\|_{\mathcal{L}(H^s(\mathbb{R}^n), H^{s-\mu}(\mathbb{R}^n))}^2 \leq \sup_{\xi \in \mathbb{R}^n} \langle\xi\rangle^{-2\mu} |f(\xi)|^2.$$

Moreover, we have

$$\kappa_{\langle\eta\rangle}^{-1} \mathrm{Op}_x(p)(y,\eta)\kappa_{\langle\eta\rangle} u(x) = \iint e^{i(\langle\eta\rangle^{-1}x - x')\cdot\xi} p(y,\xi,\eta) u(\langle\eta\rangle x)\, dx' \,\mathrm{d}\xi$$

$$= \iint e^{i(x-\widetilde{x})\cdot\langle\eta\rangle^{-1}\xi} p(y,\xi,\eta) u(\widetilde{x})\, d\widetilde{x}\, \langle\eta\rangle^{-n}\,\mathrm{d}\xi$$

$$= \iint e^{i(x-\widetilde{x})\cdot\widetilde{\xi}} p(y,\langle\eta\rangle\widetilde{\xi},\eta) u(\widetilde{x})\, d\widetilde{x}\,\mathrm{d}\widetilde{\xi}.$$

Thus, for $f(y,\widetilde{\xi},\eta) = p(y,\langle\eta\rangle\widetilde{\xi},\eta)$ from the first part of the proof we obtain

$$\left\| \kappa_{\langle\eta\rangle}^{-1} \mathrm{Op}_x(p)(y,\eta)\kappa_{\langle\eta\rangle} \right\|_{\mathcal{L}(H^s(\mathbb{R}^n), H^{s-\mu}(\mathbb{R}^n))}^2 = \sup_{\xi \in \mathbb{R}^n} \langle\xi\rangle^{-2\mu} \left| p(y,\langle\eta\rangle\xi,\eta) \right|^2$$

$$\leq c \sup_{\xi \in \mathbb{R}^n} \langle\xi\rangle^{-2\mu} \langle\langle\eta\rangle\xi,\eta\rangle^{2\mu}$$

$$\leq c\langle\eta\rangle^{2\mu}.$$

In the latter estimate we employed the relation

$$\langle\langle\eta\rangle\xi,\eta\rangle = \langle\xi\rangle\langle\eta\rangle.$$

$\square$

Corollary 1.34 *Because of*

$$D_y^\alpha D_\eta^\beta p(y, \xi, \eta) \in S^{\mu - |\beta|}(\Omega \times \mathbb{R}_{\xi, \eta}^{n+q}),$$

it follows that

$$\left\| \kappa_{\langle \eta \rangle}^{-1} \{ D_y^\alpha D_\eta^\beta \mathrm{Op}_x(p)(y, \eta) \} \kappa_{\langle \eta \rangle} \right\|_{\mathcal{L}(H^s(\mathbb{R}^n), H^{s-\mu}(\mathbb{R}^n))}^2 \leq c \langle \eta \rangle^{\mu - |\beta|}$$

for all $(y, \eta) \in M \times \mathbb{R}^n$, $M \Subset \Omega$, *and* $\alpha, \beta \in \mathbb{N}^n$, *for constants* $c = c(\alpha, \beta, M) > 0$.

Corollary 1.34 is a motivation for introducing operator-valued symbols with twisted symbolic estimates. Those play an important role in the pseudo-differential calculus on manifolds with edges and higher singularities, but also as boundary amplitude functions in boundary value problems, to be studied below.

1.4 Edge Sobolev Spaces

The spaces $H := H^s(\mathbb{R}^n)$ are one example of Hilbert spaces with group action in the following sense.

Definition 1.35 A (separable) Hilbert space H is said to be endowed with a group action $\kappa = \{\kappa_\lambda\}_{\lambda \in \mathbb{R}_+}$ if

$$\kappa_\lambda : H \to H$$

is a family of isomorphisms,

$$\kappa_\lambda \kappa_\delta = \kappa_{\lambda \delta} \qquad \text{for every} \qquad \lambda, \delta \in \mathbb{R}_+,$$

and $\lambda \to \kappa_\lambda h$ defines an element of $C(\mathbb{R}_+, H)$ for every $h \in H$.

Proposition 1.36 *Let H be a Hilbert space with group action* $\kappa = \{\kappa_\lambda\}_{\lambda \in \mathbb{R}_+}$. *Then there are constants $C, M > 0$ such that*

$$\left\| \kappa_\lambda \right\|_{\mathcal{L}(H)} \leq C (\max\{\lambda, \lambda^{-1}\})^M. \tag{1.63}$$

Proof By assumption, for every $h \in H$, the function $\lambda \to \kappa_\lambda h$ belongs to $C(\mathbb{R}_+, H)$. Thus the set $\{\kappa_\lambda h : \lambda \in [\alpha, \beta]\}$ is a bounded set in H for every compact interval $[\alpha, \beta] \subset \mathbb{R}_+$. From the Banach-Steinhaus Theorem we have $\sup_{\lambda \in [\alpha, \beta]} \|\kappa_\lambda\|_{\mathcal{L}(H)} \leq C$ for some $C = C(\alpha, \beta) > 0$. Thus, in particular, $\sup_{\lambda \in [e^{-1}, e]} \|\kappa_\lambda\|_{\mathcal{L}(H)} \leq C$, where we assume $C \geq 1$. It follows that $\|\kappa_{\lambda^n}\|_{\mathcal{L}(H)} \leq \|\kappa_\lambda\|_{\mathcal{L}(H)}^n \leq C^n$, i.e.,

$$\|\kappa_\lambda\|_{\mathcal{L}(H)} \le C^n \quad \text{for every} \quad \lambda \in [e^{-n}, e^n], \ n \in \mathbb{N}.$$

Thus, for $\lambda_n := e^n$, $n \in \mathbb{N}$, it follows that

$$\|\kappa_{\lambda_n}\|_{\mathcal{L}(H)} \le C^n = e^{n \log C} = \lambda_n^{\log C}.$$

For $n \ge 1$ and $\lambda \in [e^{n-1}, e^n]$ there is a $\delta \in [1, e]$ such that $\lambda = \delta e^{n-1} = \delta \lambda_{n-1}$. For $\lambda \in [e^{n-1}, e^n]$ we obtain

$$
\begin{aligned}
\|\kappa_\lambda\|_{\mathcal{L}(H)} &= \|\kappa_{\delta\lambda_{n-1}}\|_{\mathcal{L}(H)} \le \|\kappa_\delta\|_{\mathcal{L}(H)}\|\kappa_{\lambda_{n-1}}\|_{\mathcal{L}(H)} \le CC^{n-1} \\
&= C(\lambda_{n-1})^{\log C} \le C(\delta\lambda_{n-1})^{\log C} = C\lambda^{\log C}.
\end{aligned}
\tag{1.64}
$$

Since the right-hand side of the latter estimate is independent of n, we proved the asserted estimate for $M = \log C$, for all $\lambda \ge 1$. In a similar manner we can argue for $0 < \lambda \le 1$. In fact, for $\lambda_{-n} := e^{-n}$, $n \ge 1$, we have

$$\|\kappa_{\lambda_{-n}}\|_{\mathcal{L}(H)} \le C^n = e^{n \log C} = (\lambda_{-n}^{-1})^{\log C}.$$

Moreover, for $\lambda \in [e^{-n}, e^{-(n-1)}]$ there is a $\delta \in [e^{-1}, 1]$ such that $\lambda = \delta\lambda_{-(n-1)}$. This gives us

$$
\begin{aligned}
\|\kappa_\lambda\|_{\mathcal{L}(H)} = \|\kappa_{\delta\lambda_{-(n-1)}}\|_{\mathcal{L}(H)} &\le C(\lambda_{-(n-1)}^{-1})^{\log C} \\
&\le C((\delta\lambda_{-(n-1)})^{-1})^{\log C} = C(\lambda^{-1})^{\log C}
\end{aligned}
$$

which corresponds to the assertion for $0 < \lambda \le 1$. $\qquad\square$

Remark 1.37 Let H be a Hilbert space with group action $\kappa = \{\kappa_\lambda\}_{\lambda\in\mathbb{R}_+}$. Then there is a constant $c > 0$ such that

$$\|\kappa_{\langle\xi\rangle}^{-1}\kappa_{\langle\eta\rangle}\|_{\mathcal{L}(H)} \le c\langle\xi - \eta\rangle^M \tag{1.65}$$

for all $\xi, \eta \in \mathbb{R}^q$, where M is the constant from (1.63).

In fact, it suffices to combine (1.63) with Peetre's inequality which says that

$$(1 + |\xi - \eta|)^s \le (1 + |\eta|)^s (1 + |\xi|)^{|s|} \tag{1.66}$$

for all $\xi, \eta \in \mathbb{R}^q$, and $s \in \mathbb{R}$.

Definition 1.38 Let H and $\widetilde{H}$ be Hilbert spaces with group action $\kappa = \{\kappa_\lambda\}_{\lambda\in\mathbb{R}_+}$ and $\widetilde{\kappa} = \{\widetilde{\kappa}_\lambda\}_{\lambda\in\mathbb{R}_+}$, respectively. Then for any open $\Omega \subseteq \mathbb{R}^p$ and $\mu \in \mathbb{R}$ by

$$S^\mu(\Omega \times \mathbb{R}^q; H, \widetilde{H}) \tag{1.67}$$

we denote the set of all $a(y, \eta) \in C^\infty(\Omega \times \mathbb{R}^q, \mathcal{L}(H, \widetilde{H}))$ such that

$$\|\tilde{\kappa}_{\langle\eta\rangle}^{-1}\{D_y^\alpha D_\eta^\beta a(y, \eta)\}\kappa_{\langle\eta\rangle}\|_{\mathcal{L}(H,\widetilde{H})} \leq c\langle\eta\rangle^{\mu-|\beta|} \tag{1.68}$$

for all $(y, \eta) \in M \times \mathbb{R}^q$, $M \Subset \Omega$, $\alpha \in \mathbb{N}^p$, $\beta \in \mathbb{N}^q$, for constants $c = c(\alpha, \beta, M) > 0$.

The estimates (1.68) are also referred to as twisted symbolic estimates. Clearly for $H = \widetilde{H} = \mathbb{C}$ and $\kappa_\lambda = \mathrm{id}_\mathbb{C}$, $\tilde{\kappa}_\lambda = \mathrm{id}_\mathbb{C}$ for all $\lambda \in \mathbb{R}_+$ we recover Definition 1.2 (i). The space (1.67) depends on the choice of κ and $\tilde{\kappa}$. If necessary we write $S^\mu(\Omega \times \mathbb{R}^q; H, \widetilde{H})_{\kappa,\tilde{\kappa}}$ rather than $S^\mu(\Omega \times \mathbb{R}^q; H, \widetilde{H})$.

We also have subspaces of classical symbols. Those are based on twisted homogeneity. Let

$$S^{(\mu)}(\Omega \times (\mathbb{R}^q \setminus \{0\}); H, \widetilde{H})$$

be the set of all $a_{(\mu)}(y, \eta) \in C^\infty(\Omega \times (\mathbb{R}^q \setminus \{0\}), \mathcal{L}(H, \widetilde{H}))$ such that

$$a_{(\mu)}(y, \lambda\eta) = \lambda^\mu \tilde{\kappa}_\lambda a_{(\mu)}(y, \eta)\kappa_\lambda^{-1}$$

for all $\lambda \in \mathbb{R}_+$.

Definition 1.39 $S_{\mathrm{cl}}^\mu(\Omega \times \mathbb{R}^q; H, \widetilde{H})$ is the subspace of all $a(y, \eta) \in S^\mu(\Omega \times \mathbb{R}^q; H, \widetilde{H})$ such that there are $a_{(\mu-k)}(y, \eta) \in S^{(\mu-k)}(\Omega \times (\mathbb{R}^q \setminus \{0\}); H, \widetilde{H})$, such that for any excision function $\chi(\eta)$

$$a(y, \eta) - \sum_{k=0}^{N} \chi(\eta)a_{(\mu-k)}(y, \eta) \in S^{\mu-(N+1)}(\Omega \times \mathbb{R}^q; H, \widetilde{H})$$

for all $N \in \mathbb{N}$.

Example 2

(1) Let $\mathbf{e}^+ : L^2(\mathbb{R}_+) \to L^2(\mathbb{R})$ be the operator of extension by zero to $\mathbb{R}_-$, i.e.,

$$\mathbf{e}^+ u = \begin{cases} u & \text{on} \quad \mathbb{R}_+, \\ 0 & \text{on} \quad \mathbb{R}_-. \end{cases}$$

Then, for

$$(\kappa_\lambda u)(r) = \lambda^{\frac{1}{2}} u(\lambda r), \qquad \lambda \in \mathbb{R}_+,$$

$u \in L^2(\mathbb{R}_+)$ (or $u \in L^2(\mathbb{R})$) we have

$$\mathbf{e}^+ u \in S^{(0)}(\mathbb{R}^q; L^2(\mathbb{R}_+), L^2(\mathbb{R}))$$

for any $q \in \mathbb{N}$. At the same time we have $\mathbf{e}^+ \in S^0_{\mathrm{cl}}(\mathbb{R}^q; L^2(\mathbb{R}_+), L^2(\mathbb{R}))$.

(ii) Let $\mathbf{r}^+ : H^s(\mathbb{R}) \to H^s(\mathbb{R}_+)$ be the operator of restriction to $\mathbb{R}_+$. Then, again for $\kappa_\lambda u(r) = \lambda^{\frac{1}{2}} u(\lambda r), \lambda \in \mathbb{R}_+$, we have

$$\mathbf{r}^+ \in S^{(0)}(\mathbb{R}^q \setminus \{0\}; H^s(\mathbb{R}), H^s(\mathbb{R}_+)) \quad \text{and} \quad \mathbf{r}^+ \in S^0_{\mathrm{cl}}(\mathbb{R}^q; H^s(\mathbb{R}), H^s(\mathbb{R}_+))$$

for every $s \in \mathbb{R}$.

(iii) Let $\mathbf{r}' : H^s(\mathbb{R}) \to \mathbb{C}$ for $s > \frac{1}{2}$ be the operator of restriction $u \to u(0)$. Then for $\widetilde{\kappa}_\lambda$ on $H^s(\mathbb{R})$ as in (ii) and $\kappa_\lambda = \mathrm{id}_{\mathbb{C}}$ we have

$$\mathbf{r}' \in S^{(\frac{1}{2})}(\mathbb{R}^q \setminus \{0\}; H^s(\mathbb{R}), \mathbb{C}) \quad \text{and} \quad \mathbf{r}' \in S^{\frac{1}{2}}_{\mathrm{cl}}(\mathbb{R}^q; H^s(\mathbb{R}), \mathbb{C}).$$

More generally, the operator

$$\mathbf{r}' \frac{d^j}{dr^j} : H^s(\mathbb{R}) \to \mathbb{C} \quad \text{for} \quad s - j > \frac{1}{2}$$

defines elements

$$\mathbf{r}' \frac{d^j}{dr^j} \in S^{(j+\frac{1}{2})}(\mathbb{R}^q \setminus \{0\}; H^s(\mathbb{R}), \mathbb{C}) \quad \text{and} \quad \mathbf{r}' \frac{d^j}{dr^j} \in S^{j+\frac{1}{2}}_{\mathrm{cl}}(\mathbb{R}^q; H^s(\mathbb{R}), \mathbb{C}).$$

Remark 1.40 Example 2 shows that twisted homogeneity of some order is possible even when the mappings in question do not depend on η at all.

In our calculus we will employ many other examples of operator-valued symbols.

An important generalization concerns Fréchet spaces E, $\widetilde{E}$ rather than Hilbert spaces with group action.

A Fréchet space E written as a projective limit

$$E = \varprojlim_{j \in \mathbb{N}} E^j$$

of Hilbert spaces E^j, where we assume continuous embeddings $E^j \hookrightarrow E^0$ for all j is said to be endowed with a group action $\kappa = \{\kappa_\lambda\}_{\lambda \in \mathbb{R}_+}$ if κ is a group action on E^0 and $\kappa|_{E^j}$ a group action on E^j for every j.

Example 3 The space $\mathcal{S}(\mathbb{R}_+) := \mathcal{S}(\mathbb{R})|_{\mathbb{R}_+}$ is Fréchet as a projective limit

$$\mathcal{S}(\mathbb{R}_+) = \varprojlim_{j \in \mathbb{N}} H^{j,j}(\mathbb{R}_+)$$

where $H^{j,j}(\mathbb{R}_+) := \langle r \rangle^{-j} H^j(\mathbb{R}_+)$, $H^s(\mathbb{R}_+) := H^s(\mathbb{R})|_{\mathbb{R}_+}$, and the above-mentioned $\{\kappa_\lambda\}_{\lambda \in \mathbb{R}_+}$ is a group action on $\mathcal{S}(\mathbb{R}_+)$.

In connection with boundary value problems in a domain of dimension n with smooth boundary, locally described $\mathbb{R}^{n-1} \times \overline{\mathbb{R}}_+$, the following kind of operator-valued symbols is of interest. In the following for abbreviation we often write $q := n - 1..$

Definition 1.41 A $g(y, \eta) \in C^\infty(\mathbb{R}^q \times \mathbb{R}^q, \mathcal{L}(L^2(\mathbb{R}_+), L^2(\mathbb{R}_+)))$ is called a Green symbol of order $\mu \in \mathbb{R}$ if g and g^* (the (y, η)-wise adjoint) are symbols $g(y, \eta)$, $g^*(y, \eta) \in S_{\mathrm{cl}}^\mu(\mathbb{R}^q \times \mathbb{R}^q; L^2(\mathbb{R}_+), \mathcal{S}(\mathbb{R}_+))$. In this notation

$$S_{\mathrm{cl}}^\mu(\mathbb{R}^q \times \mathbb{R}^q; L^2(\mathbb{R}_+), \mathcal{S}(\mathbb{R}_+)) := \varprojlim_{j \in \mathbb{N}} S_{\mathrm{cl}}^\mu(\mathbb{R}^q \times \mathbb{R}^q; L^2(\mathbb{R}_+), H^{j,j}(\mathbb{R}_+)).$$

Observe that for Green symbols $g_j(y, \eta)$, $j = 0, \ldots, d$, of order $\mu - j$ we have

$$g(y, \eta) = \sum_{j=0}^{d} g_j(y, \eta) \Big(\frac{\partial}{\partial r} \Big)^j \in S_{\mathrm{cl}}^\mu(\mathbb{R}^q \times \mathbb{R}^q; H^s(\mathbb{R}_+), \mathcal{S}(\mathbb{R}_+)) \tag{1.69}$$

for every $j \in \mathbb{N}$, $s > d - \frac{1}{2}$.

Pseudo-differential boundary value problems with the transmission property at the boundary locally in the half-space $\overline{\mathbb{R}}_+^n \ni (y, r)$, $y \in \mathbb{R}^q$, $r \in \overline{\mathbb{R}}_+$, are formulated in terms of operator-valued symbols

$$a(y, \eta) := \mathbf{r}^+ \mathrm{Op}_r(p)(y, \eta) \mathbf{e}^+ + g(y, \eta) \tag{1.70}$$

where $p(x, \xi)$ is a classical symbol in $S_{\mathrm{cl}}^\mu(\overline{\mathbb{R}}_+^n \times \mathbb{R}^n)$ of order $\mu \in \mathbb{Z}$ with the transmission property at the boundary. Here we employ the splitting of variables $x = (y, r)$ and coverables $\xi = (\eta, \rho)$. The symbols (1.70) belong to $S^\mu(\mathbb{R}_y^q \times \mathbb{R}_\eta^q; H^s(\mathbb{R}_+), H^{s-\mu}(\mathbb{R}_+))$ for every $s > d - \frac{1}{2}$. Boundary conditions of trace and potential type are encoded by symbols $t(y, \eta)$, $k(y, \eta)$, $q(y, \eta)$ where $t(y, \eta) \in S_{\mathrm{cl}}^\mu(\mathbb{R}^q \times \mathbb{R}^q; H^s(\mathbb{R}_+), \mathbb{C})$, for $s > d - \frac{1}{2}$,

$$t(y, \eta)u \;=\; \sum_{j=0}^{d-1} b_j(y, \eta)\gamma^j u \tag{1.71}$$

for

$$\gamma^j u \;:=\; \left(\frac{\partial}{\partial r}\right)^j u\Big|_{r=0},$$

$$b_j(y, \eta) \in S_{\mathrm{cl}}^{\mu-j-\frac{1}{2}}(\mathbb{R}^q \times \mathbb{R}^q), \; k(y, \eta) \in S_{\mathrm{cl}}^{\mu}(\mathbb{R}^q \times \mathbb{R}^q; \mathbb{C}, \mathcal{S}(\mathbb{R}_+)),$$

and $q(y, \eta) \in S_{\mathrm{cl}}^{\mu}(\mathbb{R}^{n-1} \times \mathbb{R}^{n-1})$.

More generally operator-valued boundary symbols of pseudo-differential boundary value problems with the transmission property are block matrix symbols of the form

$$\mathbf{a}(y, \eta) \;:=\; \begin{pmatrix} a(y, \eta) & k(y, \eta) \\ t(y, \eta) & q(y, \eta) \end{pmatrix} \in S^{\mu}\big(\mathbb{R}^q \times \mathbb{R}^q; H^s(\mathbb{R}_+) \oplus \mathbb{C}^{j-}, H^{s-\mu}(\mathbb{R}_+) \oplus \mathbb{C}^{j+}\big)$$

$$\tag{1.72}$$

where $k = (k_1, \ldots, k_{j_-})$, $t = {}^{\mathrm{t}}(t_1, \ldots, t_{j_+})$ $q = (q_{ij})_{i=1,\ldots,j_+,\, j=1,\ldots,j_-}$, where the components k_i, t_j, q_{ij}, are of the above-mentioned type.

The associated operators $\mathrm{Op}_y(\cdot)$ then describe the corresponding boundary value problems themselves, locally near the boundary. Far from the boundary they contain local contributions from symbols in $S_{\mathrm{cl}}^{\mu}(\mathbb{R}^n \times \mathbb{R}^n)$. Globally the operators are described by such local operators, combined with charts and a subordinate partition of unity. This global aspect is not the main issue of the present discussion. Instead we point out the nature of $\overline{\mathbb{R}}_+^n \ni x$ as a non-compact manifold with boundary $\partial\overline{\mathbb{R}}_+^n = \mathbb{R}^{n-1} \ni x'$, and with conical exit to infinity, i.e., $|x| \to \infty$ and $|x'| \to \infty$, respectively. Similarly as in Definition 1.4 we need a variant of the operator-valued boundary symbols for an extra weight $v \in \mathbb{R}$ at infinity. For $p(x, \xi)$ with the transmission property with control of weight $v \in \mathbb{R}$ at infinity we impose the condition

$$p(x, \xi) \in S_{\mathrm{cl}}^{\mu;v}(\mathbb{R}^n \times \mathbb{R}^n)\Big|_{\overline{\mathbb{R}}_+^n \times \mathbb{R}^n}$$

and we have analogue of the three principal symbolic components (1.17), namely,

$$\sigma(p) \;=\; \big(\sigma_\psi(p),\, \sigma_{\mathrm{e}}(p),\, \sigma_{\psi,\mathrm{e}}(p)\big).$$

Similar triples appear for the operator-valued boundary symbols, namely, for (1.72)

$$\sigma'(\mathbf{a}) \;=\; \big(\sigma_\partial(\mathbf{a}),\, \sigma_{\mathrm{e}'}(\mathbf{a}),\, \sigma_{\partial,\mathrm{e}'}(\mathbf{a})\big). \tag{1.73}$$

where $\sigma_\partial(\mathbf{a})$ is the twisted homogeneous principal part of $\mathbf{a}$ in the standard sense. Moreover, in order to describe the remaining two components of (1.73), we first look at $g(y, \eta)$ in Definition 1.41. In this case we first consider (1.69) in the exit variant of weight $v \in \mathbb{R}$ by assuming $g_j(y, \eta) \in S_{\mathrm{cl}}^{\mu-j;v}(\mathbb{R}^q \times \mathbb{R}^q; H^s(\mathbb{R}_+), \mathcal{S}(\mathbb{R}_+))$, $j = 0, \ldots, d$, and then, analogously as in the scalar case we have the symbols

$$\sigma_{e'}(g_j)(y, \eta), \qquad \sigma_{\partial, e'}(g_j)(y, \eta)$$

which yield $\sigma_{e'}(g)$ and $\sigma_{\partial, e'}(g)$, respectively. For $t(y, \eta)$, $k(y, \eta)$ and $q(y, \eta)$ we have an easier relationship to notion of (1.17). First for the analogue of (1.71) we assume $b_j(y, \eta) \in S_{\mathrm{cl}}^{\mu-j-\frac{1}{2};v}(\mathbb{R}^q \times \mathbb{R}^q)$. Then according to (1.17) we have the corresponding triple of symbols, now, because of the relationship to the boundary denoted by

$$\sigma'(b_j) = \left(\sigma_\partial(b_j), \sigma_{e'}(b_j), \sigma_{\partial, e'}(b_j)\right)$$

where, for instance, $\sigma_\partial(b_j)$ is the homogeneous principal symbol of b_j in $(y, \eta) \in \mathbb{R}^q \times (\mathbb{R}^q \setminus \{0\})$ of order $\mu - j - 1/2$, etc., and then

$$\sigma'(t) = \left(\sigma_\partial(t), \sigma_{e'}(t), \sigma_{\partial, e'}(t)\right)$$

is defined by

$$\sigma_\partial(t)(y, \eta) = \sum_{j=0}^{d-1} \sigma_\partial(b_j)(y, \eta)\gamma^j,$$

$$\sigma_{e'}(t)(y, \eta) = \sum_{j=0}^{d-1} \sigma_{e'}(b_j)(y, \eta)\gamma^j,$$

$$\sigma_{\partial, e'}(t)(y, \eta) = \sum_{j=0}^{d-1} \sigma_{\partial, e'}(b_j)(y, \eta)\gamma^j.$$

Moreover, a version of the potential symbols $k(y, \eta)$ of weight $v \in \mathbb{R}$ has the form

$$k(y, \eta) \in S_{\mathrm{cl}}^{\mu;v}(\mathbb{R}^q \times \mathbb{R}^q; \mathbb{C}, \mathcal{S}(\overline{\mathbb{R}}_+))$$

where $\sigma_\partial(k)(y, \eta)$ is the twisted homogeneous principal symbol of order μ for $(y, \eta) \in \mathbb{R}^q \times (\mathbb{R}^q \setminus \{0\})$, while $\sigma_{e'}(k)(y, \eta)$ and $\sigma_{\partial, e'}(k)(y, \eta)$ are defined in an analogous manner as before. Concerning the symbol $q(y, \eta)$ in the lower right corner we assume

$$q(y, \eta) \in S_{\mathrm{cl}}^{\mu; \nu}(\mathbb{R}^q \times \mathbb{R}^q),$$

and the components coming from (1.17) are now denoted by

$$\sigma'(q) = \big(\sigma_\partial(q), \sigma_{e'}(q), \sigma_{\partial, e'}(q)\big).$$

1.5　Explicit Solutions to Dirichlet and Neumann Problems

In this section, we are going to discuss pseudo-differential calculus for boundary value problems. Let us start with a special example of an elliptic boundary value problem, namely, a Dirichlet problem

$$\mathcal{A}_1 = \begin{pmatrix} d - \Delta \\ \mathbf{r}' \end{pmatrix} : H^s(\mathbb{R}^n_+) \to \begin{array}{c} H^{s-2}(\mathbb{R}^n_+) \\ \oplus \\ H^{s-\frac{1}{2}}(\mathbb{R}^{n-1}) \end{array},$$

with a fixed constant $d > 0$. Compared with the operator $1 - \Delta$ at the very beginning we admit here $d - \Delta, d > 0$. For convenience we pass to

$$\mathcal{A}_2 = \begin{pmatrix} d - \Delta \\ Q\mathbf{r}' \end{pmatrix} : H^s(\mathbb{R}^n_+) \to \begin{array}{c} H^{s-2}(\mathbb{R}^n_+) \\ \oplus \\ H^{s-2}(\mathbb{R}^{n-1}) \end{array}, \tag{1.74}$$

where Q is an order reduction on the boundary that we take of the form $Q = \mathrm{Op}_y(\langle \eta \rangle^{\frac{3}{2}})$ such that

$$Q : H^s(\mathbb{R}^{n-1}) \to H^{s-\frac{3}{2}}(\mathbb{R}^{n-1})$$

is an isomorphism for all $s \in \mathbb{R}$.

In contrast to operators in $\mathbb{R}^n$ (or on an open subset) in boundary value problems we take into account from the very beginning additional trace operators, e.g. extra entries in a column matrix. Compositions, ellipticity, and parametrices give rise to general block 2×2 matrix operators of the form

$$\mathcal{A} = \begin{pmatrix} A + G & K \\ T & Q \end{pmatrix} : \begin{array}{c} H^s(\mathbb{R}^n_+) \\ \oplus \\ H^s(\mathbb{R}^{n-1}, \mathbb{C}^{j_1}) \end{array} \to \begin{array}{c} H^{s-\mu}(\mathbb{R}^n_+) \\ \oplus \\ H^{s-\mu}(\mathbb{R}^{n-1}, \mathbb{C}^{j_2}) \end{array}, \tag{1.75}$$

where we admit classical pseudo-differential operators A of order $\mu \in \mathbb{Z}$ with the transmission property at the boundary and trace operators T, potential operators K and

$j_2 \times j_1$ block matrices of classical pseudo-differential operators Q in the lower left corner. In addition, as we shall see below, the class of such 2×2 block matrix operators also produces so-called Green operators G in the upper left corner. Those arise in compositions and also in forming inverses or parametrices of elliptic elements.

Another essential effect is that the operators acquire from the boundary a second principal symbolic component, namely, apart from σ_ψ the so-called boundary symbol σ_∂. Finally, if we are interested in ellipticity in the half-space $\mathbb{R}^n_+$ with a control of solvability up to $|x| \to \infty$, we again observe effects from the conical exit of the configuration up to infinity. Similarly as in the introduction, this causes the presence of extra exit symbols, not only the ones in the interior, namely, σ_ψ, σ_e, $\sigma_{\psi,e}$, but also on the boundary σ_∂, $\sigma_{e'}$, $\sigma_{\partial,e'}$, as we see below. In any case, on $\overline{\mathbb{R}}^n_+$ as a special manifold with boundary, here with conical exit to infinity, we will talk about B^μ as a substitute of L^μ, used for the spaces of pseudo-differential operators in the case without boundary.

Definition 1.42 By $B^\mu(\overline{\mathbb{R}}^n_+)$ we denote the space of all operators (1.75) where the operator A in the upper left corner is an element of $L^\mu_{\mathrm{cl}}(\mathbb{R}^n_+)$ with the transmission property at the boundary, G is a Green operator, T a trace and K a potential operator in Boutet de Monvel's calculus. Concerning details we refer to [9] or [31]. For any case G and T are asked to be of type $d \in \mathbb{N}$, see the explanations below. Moreover, since $\overline{\mathbb{R}}^n_+$ is regarded as a manifold with boundary and conical exit to infinity, we assume that the symbols of the involved operators are in weighted classes, e.g., $A \in L^\mu(\mathbb{R}^n_+)$ has a symbol in $S^{\mu;\nu}_{\mathrm{cl}}(\mathbb{R}^n \times \mathbb{R}^n)|_{\overline{\mathbb{R}}^n_+ \times \mathbb{R}^n}$ for some $\nu \in \mathbb{R}$, and also the other operators in (1.75) contain the weight ν (which is kept in mind but suppressed in notation).

Then we have $\mathcal{A}_2 \in B^2(\overline{\mathbb{R}}^n_+)$ for $\nu = 0$. It is the goal of this section to show that (1.74) is an isomorphism for all $s > \frac{3}{2}$ and to construct the inverse. We have for $x = (y, r)$ and $\xi = (\eta, \rho)$

$$\sigma_\psi(\mathcal{A}_2) = |\xi|^2, \qquad \sigma_e(\mathcal{A}_2) = d + |\xi|^2, \qquad \sigma_{\psi,e}(\mathcal{A}_2) = |\xi|^2,$$

cf., notation in the introduction, in particular formula (1.5), or Remark 1.5;

$$\sigma_\partial(\mathcal{A}_2) = \begin{pmatrix} |\eta|^2 - \partial_r^2 \\ |\eta|^{\frac{3}{2}} \mathbf{r}' \end{pmatrix}, \qquad \sigma_{e'}(\mathcal{A}_2) = \begin{pmatrix} d + |\eta|^2 - \partial_r^2 \\ \langle\eta\rangle^{\frac{3}{2}} \mathbf{r}' \end{pmatrix},$$

$$\sigma_{\partial,e'}(\mathcal{A}_2) = \begin{pmatrix} |\eta|^2 - \partial_r^2 \\ |\eta|^{\frac{3}{2}} \mathbf{r}' \end{pmatrix}.$$

The latter triple of symbols coming from the boundary have been defined in general form at the end of the preceding section, here for the case $\nu = 0$. Hence $\mathcal{A}_2$ is elliptic in the sense of the symbolic triples $(\sigma_\psi, \sigma_e, \sigma_{\psi,e})$ and $(\sigma_\partial, \sigma_{e'}, \sigma_{\partial,e'})$. First we invert the operator family

$$\begin{pmatrix} \alpha^2 - \partial_r^2 \\ \beta \mathbf{r'} \end{pmatrix} : \mathcal{S}(\mathbb{R}_+) \to \begin{matrix} \mathcal{S}(\mathbb{R}_+) \\ \oplus \\ \mathbb{C} \end{matrix} \, , \tag{1.76}$$

where $\alpha^2 := (d + |\eta|^2)^{\frac{1}{2}}$, $\beta := \langle \eta \rangle^{\frac{3}{2}}$. Let us write

$$\ell_{\pm}(\rho) = \alpha \pm i\rho.$$

Then we have $\ell_-(\rho)\ell_+(\rho) = \alpha^2 + \rho^2$ and

$$\alpha^2 - \partial_r^2 = \mathrm{Op}^+(\ell_-\ell_+) = \mathrm{Op}^+(\ell_-)\mathrm{Op}^+(\ell_+)$$

where

$$\mathrm{Op}^+(b) := \mathbf{r}^+ \mathrm{Op}_r(b)\mathbf{e}^+$$

for any symbol $b(r, \rho) \in S^\mu(\mathbb{R} \times \mathbb{R})$. The latter identity is true since $\ell_-(\rho)$ is a minus function;

$$\mathrm{Op}^+(\ell_-) : \mathcal{S}(\mathbb{R}_+) \to \mathcal{S}(\mathbb{R}_+)$$

is an isomorphism. Thus in order to invert (1.76), it suffices to consider

$$\begin{pmatrix} \mathrm{Op}^+(\ell_+) \\ \beta \mathbf{r'} \end{pmatrix} : \mathcal{S}(\mathbb{R}_+) \to \begin{matrix} \mathcal{S}(\mathbb{R}_+) \\ \oplus \\ \mathbb{C} \end{matrix}$$

which is an isomorphism, because $\mathrm{Op}^+(\ell_+) : \mathcal{S}(\mathbb{R}_+) \to \mathcal{S}(\mathbb{R}_+)$ is surjective, and $\beta \mathbf{r'}$ induces an isomorphism of kernel: $\ker\left(\mathrm{Op}^+(\ell_+)\right) = \{\gamma e^{\alpha r} : \gamma \in \mathbb{C}\}$ to $\mathbb{C}$.

Now let us form the potential $v = v(\alpha) : \mathbb{C} \to \mathcal{S}(\mathbb{R}_+)$, defined by $v\gamma = \gamma\beta^{-1}e^{-\alpha r}$, $\gamma \in \mathbb{C}$. Then

$$\begin{pmatrix} \mathrm{Op}^+(\ell_+) \\ \beta \mathbf{r'} \end{pmatrix} \left(\mathrm{Op}^+(\ell_+^{-1}) \; v\right) = \begin{pmatrix} 1 \; 0 \\ 0 \; 1 \end{pmatrix},$$

because of $\mathbf{r'} \circ \mathrm{Op}^+(\ell_+^{-1}) = 0$ and $\mathrm{Op}^+(l_+)\mathrm{Op}^+(l_+)^{-1} = 1$, since $\mathrm{Op}^+(l_+)$ is a differential operator, i.e., $l_+(\rho)$ a polynomial in ρ and hence a minus symbol. Thus

$$\begin{pmatrix} \mathrm{Op}^+(\ell_+) \\ \beta\mathbf{r}' \end{pmatrix}^{-1} = \begin{pmatrix} \mathrm{Op}^+(\ell_+^{-1}) \; v \end{pmatrix}.$$

Consider now $a(\rho) = \alpha^2 + \rho^2$. The operator in (1.76) can be written in the form

$$\begin{pmatrix} \mathrm{Op}^+(a) \\ \beta\mathbf{r}' \end{pmatrix} = \begin{pmatrix} \mathrm{Op}^+(\ell_-) & 0 \\ 0 & 1 \end{pmatrix} \begin{pmatrix} \mathrm{Op}^+(\ell_+) \\ \beta\mathbf{r}' \end{pmatrix}.$$

This implies

$$\begin{pmatrix} \mathrm{Op}^+(a) \\ \beta\mathbf{r}' \end{pmatrix}^{-1} = \begin{pmatrix} \mathrm{Op}^+(\ell_+^{-1}) \; v \end{pmatrix} \begin{pmatrix} \mathrm{Op}^+(\ell_-^{-1}) & 0 \\ 0 & 1 \end{pmatrix}$$

$$= \begin{pmatrix} \mathrm{Op}^+(\ell_+^{-1})\mathrm{Op}^+(\ell_-^{-1}) \; v \end{pmatrix},$$

and we have

$$\mathrm{Op}^+(\ell_+^{-1})\mathrm{Op}^+(\ell_-^{-1}) = \mathrm{Op}^+(a^{-1}) + g$$

with a certain Green's operator g. We obtain for the inverse of (1.76)

$$\begin{pmatrix} \mathrm{Op}^+(a) \\ \beta\mathbf{r}' \end{pmatrix}^{-1} = \begin{pmatrix} \mathrm{Op}^+(a^{-1}) + g \; v \end{pmatrix}.$$

Inserting now the expressions for $\alpha = \alpha(\eta)$, $\beta = \beta(\eta)$, we easily see that the ingredients of

$$\sigma_{e'}(\mathcal{A}_2)^{-1}(\eta) = \begin{pmatrix} \mathrm{Op}^+(a^{-1})(\eta) + g(\eta) \; v(\eta) \end{pmatrix} \tag{1.77}$$

belong to the symbolic structure of $B^{-2}(\overline{\mathbb{R}}_+^n)$ for $v = 0$ (they are, of course, independent of y), and we obviously have

$$\mathcal{A}_2^{-1} = \mathrm{Op}(\sigma_{e'}(\mathcal{A}_2)^{-1}) =: \begin{pmatrix} P \; K \end{pmatrix} \in B^{-2}(\overline{\mathbb{R}}_+^n).$$

The method of calculating (1.77) gives us explicit expressions also for $\sigma_\partial(\mathcal{A}_2)^{-1}$ and $\sigma_{\partial,e'}(\mathcal{A}_2)^{-1}$, where

$$\sigma(\mathcal{A}_2)^{-1} = \begin{pmatrix} |\xi|^{-2}, (d + |\xi|^2)^{-1}, |\xi|^{-2}; \sigma_\partial(\mathcal{A}_2)^{-1}, \sigma_{e'}(\mathcal{A}_2)^{-1}, \sigma_{\partial,e'}(\mathcal{A}_2)^{-1} \end{pmatrix}.$$

Finally, we obtain

$$\mathcal{A}_1^{-1} = \left(P \, K Q^{-1} \right) \in B^{-2}(\overline{\mathbb{R}}_+^n).$$

Remark 1.43 Similar arguments apply to the Neumann problem for $d - \Delta$ in Ω, with $\mathbf{r}'\partial_r$ in place of $\mathbf{r}'$. To obtain an element with unified orders we can pass to the boundary operator $R\mathbf{r}'\partial_r$ for $R = \mathrm{Op}_y(\langle\eta\rangle^{\frac{1}{2}})$. We see that

$$\begin{pmatrix} d - \Delta \\ R\mathbf{r}'\partial_r \end{pmatrix} \in B^2(\overline{\mathbb{R}}_+^n) \tag{1.78}$$

is also elliptic in the sense of the symbolic triples

$$(\sigma_\psi, \sigma_e, \sigma_{\psi,e}) \quad \text{and} \quad (\sigma_\partial, \sigma_{e'}, \sigma_{\partial,e'}),$$

and even invertible as an operator $H^s(\mathbb{R}_+^n) \to H^{s-2}(\mathbb{R}_+^n) \oplus H^{s-2}(\mathbb{R}_+^n)$ for $s > \frac{3}{2}$. The inverse belongs to $B^{-2}(\overline{\mathbb{R}}_+^n)$.

Remark 1.44 Of course we may discuss the Dirichlet and Neumann problems also on a bounded domain $\Omega \Subset \mathbb{R}^n$ with smooth boundary $\partial\Omega$. Then the aspects at conical exits to infinity disappear.

1.6   Hölder and L^p Estimates for Coercive Boundary Value Problems

Let Ω be a relatively compact domain in $\mathbb{R}^{n+1}$ with smooth boundary $b\Omega$, and consider the boundary value problem

$$\begin{cases} P(x, D) = 0 \\ Q_j(x, D)u|_{b\Omega} = g_j, \ j = 1, \dots, m \end{cases} \tag{1.79}$$

where $P(x, D)$ is a strongly elliptic differential operator of order $2m$ with coefficients smooth up to the boundary, and g_j are given functions on the boundary (the case $P(x, D)u = f$ with f given on Ω can be reduced to the case $f = 0$ by use of a parametrix). Then under certain conditions called coercive, Lopatinski-Shapiro, or Agmon-Douglis-Nirenberg by various authors (cf. L. Hörmander [33]) operator A_j mapping $C^\infty(b\Omega)$ to $C^\infty(\overline{\Omega})$ can be constructed such that if u satisfies (1.79) then

$$u = \sum_{j=1}^{m} A_j g_j + Su \tag{1.80}$$

where S is a smoothing operator (cf. A.P. Calderón [10], L. Hörmander [34], L. Boutet de Monvel [9]). The operators A_j which play an analogous role to the Poisson kernel in the case of the Dirichlet problem for Laplacian can be described more precisely as follows: using appropriate coordinate charts and partitions of unity, we can cover $\overline{\Omega}$ by an open set $\in \Omega$ and a system of open set of $\overline{\Omega}$ which are diffeomorphic to open sets of $\overline{\mathbb{R}^{n+1}_+}$, on the interior open set the A_j's are infinitely smoothing, while on the others they are pseudo-Poisson operators in the sense given below:

Definition 1.45 A function $a(x, y, \xi) \in C^{\infty}(\mathbb{R}^n \times \mathbb{R}_+ \times \mathbb{R}^n)$ is said to be a pseudo-Poisson symbol of order m if it is compactly supported in x and y and if there exists constant C (independent of x, y, ξ and depending perhaps on α, β, γ, δ) such that

$$|y^{\alpha} D_y^{\beta} D_x^{\gamma} D_{\xi}^{\delta} a(x, y, \xi)| \leq C \langle \xi \rangle^{m - |\alpha| + |\beta| - |\delta|} \tag{1.81}$$

The operator mapping on $C_0^{\infty}(\mathbb{R}^n)$ to $C_0^{\infty}(\overline{\mathbb{R}^{n+1}_+})$ defined by

$$Au(x, y) = \int e^{ix \cdot \xi} a(x, y, \xi) \widehat{u}(\xi) d\xi \tag{1.82}$$

is then said to be a pseudo-Poisson operator of order m.

In (1.80) A_j will have $- m_j$ if $Q_j(x, D)$ has order m_j as a differential operator. Thus, estimate and regularity theorems for coercive boundary value problems reduce to boundedness properties of pseudo-Poisson operators. To state these we first need some definitions:

J_{α}^n = pseudo-differential operator on $\mathbb{R}^n$ with symbol $\langle \xi \rangle^{\frac{\alpha}{2}}$ (Bessel potential of order α).
$L_k^p(\mathbb{R}^{n+1})$ = space of functions $f \in L^p(\mathbb{R}^{n+1})$ such that $J_k^{n+1} f \in L^p(\mathbb{R}^{n+1})$.
$L_k^p(\mathbb{R}^{n+1}_+)$ = space of restrictions to $\mathbb{R}^{n+1}_+$ of functions in $L_k^p(\mathbb{R}^{n+1})$.
$B^p(\mathbb{R}^n)$ = space of all $f \in L^p(\mathbb{R}^n)$, $f(x) = F(x, 0)$ for some function $F \in L_1^p(\mathbb{R}^{n+1})$.

   The norm on B^p will be $\|f\|_{B^p} = \inf \|F\|_{L_1^p(\mathbb{R}^{n+1})}$.
$\Lambda_{\alpha}(\mathbb{R}^n)$ = space of all function Hölder of class α in $\mathbb{R}^n$.
$\Lambda_{\alpha}(\mathbb{R}^{n+1}_+)$ = space of all function Hölder of class α in $\mathbb{R}^{n+1}_+$.

Henceforth, unless otherwise stated we shall assume that all pseudo-differential operators have symbols $a(x, \xi)$ compactly supported in x. The main results of this section in terms of isotropic norms are summarized in the following:

Theorem 1.46 *Let K denote a compact subset of $\mathbb{R}^n$. Then, any pseudo-Poisson operator of order 0 can be extended as a bounded operator from*

(1) $\varepsilon'(K) \cap \Lambda_\alpha(\mathbb{R}^n)$ *to* $\Lambda_\alpha(\mathbb{R}^{n+1}_+)$

(2) $\varepsilon'(K) \cap B^p(\mathbb{R}^n)$ *to* $L^p_1(\mathbb{R}^{n+1}_+)$

(3) $\varepsilon'(K) \cap L^2(\mathbb{R}^n)$ *to* $H_{(1,-\frac{1}{2})}(\mathbb{R}^{n+1}_+)$

   Furthermore, any pseudo-differential operator of order m can be extended boundedly from

(4) $\varepsilon'(K) \cap \Lambda_\alpha(\mathbb{R}^n)$ *to* $\Lambda^p_{\alpha-m}(\mathbb{R}^n)\ 0 < \alpha < \infty$

(5) $\varepsilon'(K) \cap L^p_k(\mathbb{R}^n)$ *to* $L^p_{k-m}(\mathbb{R}^n)\ 1 < p < \infty$

(6) $\varepsilon'(K) \cap B^p(\mathbb{R}^n)$ *to* $B^p(\mathbb{R}^n)$ *when* $m = 0,\ 1 < p < \infty$

Proof of (1) Assume first that $0 < \alpha < 1$. For each fixed $y > 0$ then the pseudo-differential operator A_y defined by $A_y u(x) = Au(x, y)$ is of order $-\infty$, and consequently $Au(x, y) \in C^\infty(\mathbb{R}^{n+1}_+)$. According to a classical theorem of Hardy and Littlewood, it thus suffices to show that

$$|\nabla_{x,y} Au(x, y)| \leq C \|u\|_{\Lambda_\alpha} y^{\alpha-1} \tag{1.83}$$

$\Leftrightarrow Au \in \Lambda_\alpha(\mathbb{R}^{n+1})$. In this case, $\delta(x, y) = y$, where $\delta(x, y)$ is the distance from (x, y) to the boundary.

   Now the kernel $K(x, y, z)$ of A_y is given by the integral

$$K(x, y, z) = \int e^{-i\xi \cdot z} a(x, y, \xi) d\xi \tag{1.84}$$

which converges absolutely since $a(x, y, \xi)$ is rapidly decreasing in ξ for each (x, y); thus (1.82) can be rewritten as

$$Au(x, y) = \int e^{ix \cdot \xi} a(x, y, \xi) \widehat{u}(\xi) d\!\!\!/\xi$$

$$= \iint e^{i(x-z) \cdot \xi} a(x, y, \xi) u(z) dz d\!\!\!/\xi$$

$$= \int K(x, y, x - z) u(z) dz$$

and hence

$$\nabla_y Au(x, y) = \int \frac{\partial K}{\partial y}(x, y, x - z) u(z) dz \tag{1.85}$$

$$\nabla_x Au(x, y) = \int \left(\frac{\partial K}{\partial x}(x, y, x - z) + \frac{\partial K}{\partial z}(x, y, x - z)\right)u(z)dz \tag{1.86}$$

To estimate the integrals in (1.83) and (1.84) we first determine the degree of singularity of $K(x, y)$ near the origin and infinity. $\square$

Lemma 1.47 *There exist constants $C_{\alpha,\beta,\gamma,\delta} > 0$ such that for all x, y, and z*

$$|y^\delta D_x^\alpha D_y^\beta D_z^\gamma K(x, y, z)| \le C_{\alpha,\beta,\gamma,\delta}|z|^{-n-|\beta|-|\gamma|+|\delta|} \tag{1.87}$$

Proof Since kernels corresponding to symbols compactly supported in ξ will be Schwartz functions of z with bounds uniform with respect to both x and y and hence will trivially satisfy (1.87), we may assume that $a(x, y, \xi)$ vanishes for $|\xi| \le 1$. Let $\psi(\theta) \in C_0^\infty(\mathbb{R}^n)$ be a function which is identically 0 for $|\theta| \le 1$ and 1 for $|\theta| \ge 2$. Using ψ to decompose (1.84) into two integrals and integrating by parts in the second one yields

$$K(x, y, z) = \int e^{-2\pi i\xi\cdot z}a(x, y, \xi)d\xi$$

$$= \int e^{-i\xi\cdot z}a(x, y, \xi)(1 - \psi(|z|\xi))d\xi$$

$$+ \int e^{-i\xi\cdot z}a(x, y, \xi)\psi(|z|\xi)d\xi$$

$$y^\delta D_x^\alpha D_y^\beta D_z^\gamma K(x, y, z) = \int e^{i\xi\cdot z}y^\delta\xi^\gamma D_x^\alpha D_y^\beta a(x, y, \xi)(1 - \psi(|z|\xi))d\xi$$

$$+ |z|^{-2N}\int e^{-i\xi\cdot z}\Delta_\xi^N(y^\delta\psi(|z|\xi)\xi^\gamma D_x^\alpha D_y^\beta a(x, y, \xi))d\xi$$

$$= I_1 + |z|^{-2N}I_2$$

Since (1.81) together with the fact $a(x, y, \xi) = 0$ when $|\xi| \le 1$ imply

$$|y^\delta\xi^\gamma D_x^\alpha D_y^\beta a(x, y, \xi)| \le C(\alpha, \beta, \gamma, \delta)|\xi|^{|\beta|+|\gamma|-\delta}$$

it is easy to see that $|I_1|$ is bounded by $C|z|^{-n-|\beta|-|\gamma|+\delta}$; as for I_2

$$|I_2| \le C \sum_{|k|+|\ell|=2N}\int |z|^\ell(D_\xi^k\psi(|z|\xi))|D_\xi^\ell(y^\delta\xi^\gamma D_x^\alpha)D_y^\beta a(x, y, \xi)|d\xi$$

$$\le C|z|^{2N-n-|\beta|-|\gamma|+\delta}$$

and the proof of Lemma 1.47 is complete. $\square$

We now estimate (1.85) and (1.86) as follows : Write

$$\nabla_y Au(x,y) = \int \frac{\partial K}{\partial y}(x,y,x-z)(u(z)-u(x))dz + u(x)\int \frac{\partial K}{\partial y}(x,y,x-z)dx$$

$$I = \int\limits_{|x-z|\leq y} + \int\limits_{|x-z|\geq y} = I_1 + I_2$$

$$|I_1| \leq \|u\|_{\Lambda_\alpha} \int\limits_{|x-z|\leq y} |\frac{\partial K}{\partial y}(x,y,x-z)| \cdot |x-z|^\alpha dz$$

$$\leq C\|u\|_{\Lambda_\alpha} y^{-1} \int\limits_{|x-z|\leq y} |x-z|^{-n+\alpha} dz$$

$$= C\|u\|_{\Lambda_\alpha} y^{\alpha-1}$$

$$|I_2| \leq C\|u\|_{\Lambda_\alpha} \int\limits_{|x-z|\geq y} |x-z|^{-(n+1)+\alpha} dz$$

$$= C\|u\|_{\Lambda_\alpha} y^{\alpha-1}$$

On the other hand

$$\int \frac{\partial K}{\partial y}(x,y,x-z)dz = \frac{\partial a}{\partial y}(x,y,0)$$

in view of the Fourier inversion formula; I_2 is thus bounded by $C\|u\|_{L^\infty}$, and the desired estimate (1.87) follows. $\nabla_x Au(x,y)$ can be estimated in a similar way, completing the proof of the case $0 < \alpha < 1$. $\square$

When $\alpha > 1$, α non-integer, observe that

$$D_{x_j} a(x,y,\xi)\frac{\xi_j}{|\xi|^2} \quad \text{and} \quad D_y a(x,y,\xi)\frac{\xi_j}{|\xi|^2}$$

are all pseudo-Poisson symbols of order 0 (here again we have assumed that $a(x,y,\xi)$ vanishes for $|\xi| \leq 1$, which is no loss of generality). We can thus apply the previous results (with Λ_α replaced by $\Lambda_{\alpha-[\alpha]}$) to the derivatives $D^\beta u$, $|\beta| \leq [\alpha]$ and see that A can still be extended boundedly in Λ_α. Finally the case $\alpha = 1$ follows from Lions-Peetre interpolation between the case $\alpha = \frac{1}{2}$ and $\alpha = \frac{3}{2}$, and the general case of α non-negative integer is obtained from $\alpha = 1$ by the same argument as above.

Proof of (2) We shall reduce the problem to proving the L^p boundedness of certain classes of pseudo-differential operators on $\mathbb{R}^{n+1}$. Let $a(x,y,\xi) \in C^\infty(\mathbb{R}^n \times \mathbb{R}_+ \times \mathbb{R}^n)$ be pseudo-Poisson symbol of order 1 and define

$$a^{\#}(x, \eta, \xi) = \int_0^\infty \sin(yn)a(x, y, \xi)dy, \ \eta \in \mathbb{R}.$$

$\square$

Lemma 1.48 *There exists a constant $C > 0$ such that for all $(x, y, \xi) \in \mathbb{R}^{2n+1}$*

$$|a^{\#}(a, \eta, \xi)| = |\int_0^\infty \sin(yn)a(x, y, \xi)dy| \le C.$$

Proof Let $\psi(\theta) \in C_0^\infty(\mathbb{R})$ be a function which vanishes for $|\theta| \le 1$ and equals 1 for $|\theta| \ge 2$. Then

$$a^{\#}(x, y, \xi) = \int_0^\infty \sin <\eta, y> (1 - \psi(y|\eta|))a(x, y, \xi)dy$$

$$+ (-1)^N |\eta|^{-2N} \int_0^\infty \sin <\eta, y> (\frac{\partial}{\partial y})^{2N}(\psi(y|\eta|)a(x, y, \xi))dy$$

$$= I_1 + |\eta|^{-2N} I_2.$$

Since $|\sin <\eta, y>| \le |\eta|y$ and $|ya(x, y, \xi)| \le C$ we obviously have $|I_1| \le C$; as for the second integral Leibniz formula show that

$$|I_2| \le C \sum_{k=0}^{2N} \int_0^\infty |\eta|^k (D_y^k \psi(y|\eta|)) D_y^{2N-k} a(x, y, \xi)dy$$

$$\le C \sum_{k=0}^{2N} |\eta|^k \int_0^\infty y^{-2N+k-1} |D_y^k \psi(y|\eta|)| \cdot |y^{2N-k+1} D_y^{2N-k} a(x, y, \xi)|dy$$

$$\le C \sum_{k=0}^{2N} |\eta|^k \int_0^\infty y^{-2N+k-1} |D_y^k \psi(y|\eta|)|dy$$

$$\le C |\eta|^{2N}$$

as was to be proved. $\square$

Lemma 1.49 *There exists constants $C_{\alpha,\beta,\gamma} > 0$ such that for all $(x, \eta, \xi) \in \mathbb{R}^{2n+1}$*

$$|D_x^\gamma D_\eta^\beta D_\xi^\alpha a^{\#}(x, \eta, \xi)| \le C_{\alpha,\beta,\gamma}(1 + |\eta|)^{-|\beta|}(1 + |\xi|)^{-|\alpha|} \ \textit{for } |\beta| \le 1.$$

Proof When $\beta = 0$ the above inequality follows immediately from Lemma 1.47. applied to the pseudo-symbol $(1 + |\xi|^2)^{\frac{|\alpha|}{2}} D_x^\gamma D_\xi^\alpha a(x, y, \xi)$. In proving the case for $\beta = 1$ we shall assume for the sake convenience of notation that $|\alpha| = |\gamma| = 0$. It will be obvious that the proof carries over to the general case. Now

$$\eta D_\eta a^\#(x, y, \xi) = \int_0^\infty \eta y \cos < y, \eta > a(a, y, \xi) dy$$

$$= \int_0^\infty \frac{\partial}{\partial y}(\sin < y, \eta >)(ya(a, y, \xi)) dy$$

$$= -\int_0^\infty \sin < y, \eta > \frac{\partial}{\partial y}(ya(x, y, \xi)) dy$$

after an integration by parts where the boundary terms vanish; since $\frac{\partial}{\partial y}(ya(x, y, \xi))$ is pseudo-Poisson symbol of order 1, we can again apply Lemmas 1.47 and 1.48. to obtain the desired inequality. $\qquad\qquad\square$

Lemma 1.50 *The operator $A^\# : C_0^\infty(\mathbb{R}^{n+1}) \to C^\infty(\mathbb{R}^{n+1})$ defined by*

$$A^\# v(x, y) = \iint e^{ix\cdot\xi} e^{iy\cdot\eta} a^\#(x, \eta, \xi) \widehat{v}(\xi, \eta) đ\xi đ\eta \qquad (1.88)$$

can be extended as a bounded operator from $L^p(\mathbb{R}^{n+1})$ to $L^p(\mathbb{R}^{n+1})$. Here $\widehat{v}(\xi, \eta)$ denotes the Fourier transform of $v(x, y)$ with respect to both variables x and y.

Proof Let

$$b(\lambda, \eta, \xi) = \int e^{-i<\lambda,x>} a^\#(x, \eta, \xi) dx$$

and rewrite $A^\#$ as

$$A^\# v(x, y) = \int e^{i\lambda\cdot x}[\iint e^{ix\cdot\eta} e^{iy\cdot\eta} b(\lambda, \eta, \xi) \widehat{v}(\xi, \eta) đ\xi đ\eta] đ\lambda \qquad (1.89)$$

In view of Lemma 2.4. we have

$$|D_\eta^\alpha D_\xi^\beta b(\lambda, \eta, \xi)| \leq C_N(1 + |\lambda|)^{-N}(1 + |\eta|)^{-|\alpha|}(1 + |\xi|)^\beta \ |\alpha| \leq 1$$

In particular, if $\zeta \in \mathbb{R}^{n+1}$ denotes (ξ, η) then

$$|D_{\zeta_{j_1}} \ldots D_{\zeta_{j_k}} b(\lambda, \eta, \xi)| \leq C_N(1 + |\lambda|)^{-N} \prod_{\ell=1}^{k}(1 + |\zeta_{j\ell}|)^{-1}$$

where the $\{j\ell\}_{\ell=1}^k$ are all distinct; applying the non-isotropic form of the Marcinkiewicz multiplier theorem (cf. [80] Chapters VI and VII) to the operator defined by the inner integral in (1.89) and integrating with respect to λ yields

$$\|A^{\#}v(x, y)\|_{L^p(\mathbb{R}^{n+1})} \leq C\|v\|_{L^p(\mathbb{R}^{n+1})}$$

and the lemma follows immediately. $\square$

We now return to the *Proof of (2)*. That $\|Au\|_{L^p(\mathbb{R}^{n+1}_+)} \leq C\|u\|_{L^p(\mathbb{R}^n)}$ is trivial; to prove that $\|D_{x,y}Au(x, y)\|_{L^p(\mathbb{R}^{n+1}_+)} \leq C\|u\|_{B^p(\mathbb{R}^n)}$ we shall express $D_{x,y}Au(x, y)$ as restrictions of operators of the type introduced in Definition 1.45. Let $u(x) \in C_0^\infty(\mathbb{R}^n)$ and let $v(x, y) \in C_0^\infty(\mathbb{R}^{n+1})$ be a function such that

$$v(x, 0) = u(x) \text{ and } \|v\|_{L^p_1(\mathbb{R}^{n+1})} \leq 2\|u\|_{B^p(\mathbb{R}^n)}.$$

Integrating the identity

$$\frac{\partial}{\partial\lambda}(a(x, y+\lambda, \xi)\widetilde{v}(\xi, \lambda)) = \frac{\partial a}{\partial\lambda}(x, y+\lambda, \xi)\widetilde{v}(\xi, \lambda) + a(x, y+\lambda, \xi)\frac{\partial\widetilde{v}}{\partial\lambda}$$

between 0 and ∞ yields

$$a(x, y, \xi)\widehat{u}(\xi) = \int_0^\infty \frac{\partial a}{\partial\lambda}(x, y+\lambda, \xi)\widehat{v}(\xi, \lambda)d\lambda + \int_0^\infty a(x, y+\lambda, \xi)\frac{\partial\widetilde{v}}{\partial\lambda}(\xi, \lambda)d\lambda$$

and hence

$$Au(x, y) = \iint_0^\infty e^{ix\cdot\xi}\frac{\partial a}{\partial\lambda}(x, y+\lambda, \xi)\widehat{v}(\xi, \lambda)d\lambda d\xi$$

$$+ \iint_0^\infty e^{ix\cdot\xi}a(x, y+\lambda, \xi)\frac{\partial\widetilde{v}}{\partial\lambda}(\xi, \lambda)d\lambda d\xi$$

$\square$

Define then an operator to be an extended Poisson operator if it can be written in the form

$$Pv(x, y) = \iint_0^\infty e^{ix\cdot\xi}a(x, y+\lambda, \xi)\widetilde{v}(\xi, \lambda)d\lambda d\xi \quad v \in C_0^\infty(\mathbb{R}^{n+1})$$

where $a(x, y, \xi)$ is a pseudo-Poisson symbol of order 1. It is obvious from (1.87) that $D_{x_j}Au(x, y)$ is the sum of extended Poisson operators acting on the functions $D_{x_j}v(x, y)$ and $D_y v(x, y)$, since $\frac{\partial^2 a}{\partial\lambda^2}(x, y+\lambda, \xi)$ can written as

$$\frac{\partial^2 a}{\partial\lambda^2}(x, y+\lambda, \xi) = \sum_{j=1}^n \xi_j \frac{\xi_j}{|\xi|^2}a(x, y+\lambda, \xi)$$

it follows that the same holds for $D_y Au(x, y)$, and the *Proof of (2)* will be complete once the following lemma is established:

Lemma 1.51 *If P is an extended-Poisson operator then there exists some constant $C > 0$ such that for all functions $v \in C_0^\infty(\mathbb{R}^{n+1})$*

$$\|Pv(x, y)\|_{L^p(\mathbb{R}_+^{n+1})} \le C\|v\|_{L^p(\mathbb{R}^{n+1})} \tag{1.90}$$

Proof Let $w(x, y)$ be the function which equals $v(x, y)$ for $y \ge 0$ and vanishes identically for $y < 0$. It will of course suffice to prove (1.90) with $\|w\|_{L^p}$ instead of $\|v\|_{L^p}$ on the right-hand side. If we extend $a(x, y, \xi)$ to be odd as a function of y for each (x, ξ) then $Pv(x, y)$ appears to be the restriction on $\mathbb{R}_+^{n+1}$ of the operator on $\mathbb{R}^{n+1}$

$$P^\# w(x, y) = \iint_{-\infty}^{\infty} e^{ix\cdot\xi} a(x, y + \lambda, \xi)\widetilde{w}(\xi, \lambda)d\lambda \text{\dj}\xi \tag{1.91}$$

and thus we need only show

$$\|P^\# w(x, y)\|_{L^p(\mathbb{R}_+^{n+1})} \le C\|w\|_{L^p(\mathbb{R}_+^{n+1})} \tag{1.92}$$

for all functions w arising from functions $v \in C_0^\infty(\mathbb{R}^{n+1})$ by truncation as above. Call the space of all such functions a. We now observe that (1.90) would follow if

$$\|P^\# w(x, y)\|_{L^p(\mathbb{R}^{n+1})} \le C\|w\|_{L^p(\mathbb{R}^{n+1})} \tag{1.93}$$

were true for all functions $w \in C_0^\infty(\mathbb{R}^{n+1})$; indeed the operator given by formula (1.91) on $C_0^\infty(\mathbb{R}^{n+1})$ functions would then extend boundedly to $L^p(\mathbb{R}^{n+1})$ in a unique way, and restricted to functions w in a would coincide with $P^\# w(x, y)$ for $y > 0$. This can be seen by approximating any given function w in a by a sequence $\{w_n\}$ of functions in $C_0^\infty(\mathbb{R}^{n+1})$ with support in $\{(x, y) \in \mathbb{R}^{n+1} ; y > -\frac{1}{n}\}$.

Now according to the Fourier inversion formula, if $w \in C_0^\infty(\mathbb{R}^{n+1})$

$$P^\# w(x, y) = \iint_{-\infty}^{\infty} e^{ix\cdot\xi} a(x, y + \lambda, \xi) \int_{-\infty}^{\infty} e^{i\lambda\cdot\eta}\widehat{w}(\xi, \eta)\text{\dj}\eta\, d\lambda\,\text{\dj}\xi$$

$$= \iint e^{ix\cdot\xi} e^{-iy\cdot\eta}[\int_{-\infty}^{\infty} e^{i(y+\lambda)\cdot\eta} a(x, y + \lambda, \xi)d\lambda]\widehat{w}(\xi, \eta)\text{\dj}\xi\,\text{\dj}\eta$$

$$= \iint e^{ix\cdot\xi} e^{-iy\cdot\eta} a^\#(x, \eta, \xi)\widehat{w}(\xi, \eta))\text{\dj}\xi\,\text{\dj}\eta$$

which is an operator of the type (1.88), and therefore (1.93) is a consequence of Lemma 1.50. $\qquad\qquad\square$

Proof of (3) Since any pseudo-Poisson operator A of order 0 maps $L^2(\mathbb{R}^n)$ to $H_{\frac{1}{2}}(\mathbb{R}^{n+1}_+)$ in view of statement (2) of Theorem 1.46, the desired conclusion is equivalent to the boundedness of $D_y A$ from L^2 to $H_{(0,-\frac{1}{2})}$. Thus it is enough to show that if P is any pseudo-Poisson operator of order 1 then

$$|\textstyle\iint (1+|\eta|^2)^{-\frac{1}{2}}|(Pu)^\sim(\eta,y)|^2 d\eta dy| \le C\|u\|^2_{(0)},\ u \in C^\infty_0(\mathbb{R}^n) \tag{1.94}$$

C being a constant independent from u. $\qquad\qquad\qquad\qquad\qquad\qquad\qquad\square$

We first assume that the symbol a of P does not depend on x.

Lemma 1.52 *There exists a constant $C > 0$ independent from η such that*

$$\int |a(y,\eta)|^2 dy \le C|\eta| \ \textit{for}\ |\eta| \ge 1.$$

Proof

$$\int |a(y,\eta)|^2 dy = \int\limits_{y\le|\eta|^{-1}} + \int\limits_{y\ge|\eta|^{-1}} = I_1 + I_2$$

$$|I_1| \le \int\limits_{y\le|\eta|^{-1}} |\eta|^2 dy \le |\eta|$$

$$|I_2| \le \int\limits_{y\ge|\eta|^{-1}} y^{-2}|ya(y,\eta)|^2 dy \le C\int_{|\eta|-1}^\infty y^{-2}dy = C|\eta|.$$

$$\square$$

Since $a(y,\xi)$ does not depend on x, $(Pu)^\sim(\eta,y)$ is obviously equal to $a(y,\eta)\hat{u}(\eta)$, and (1.90) follows from Lemma 1.52 and the Plancherel formula. The general case of $a(x,y,\xi)$ depending on and compactly supported in x can be reduce to the previous case as follows: let

$$b(\lambda,y,\xi) = \int e^{-i\lambda\cdot x} a(x,y,\xi) dx$$

and write

$$Pu(x,y) = \int e^{i\lambda\cdot x}\int e^{ix\cdot\xi} b(\lambda,y,\xi)\hat{u}(\xi)\,d\xi\,d\lambda$$

$$= \int e^{i\lambda\cdot x} P_\lambda u\,d\lambda$$

with obvious notations; an inspection of the preceding proof shows that P_λ is bounded from L^2 to $H_{(0,\frac{1}{2})}$ with norm $\|P_\lambda\| \leq C \sup_\eta |\eta|^{-1} \int |b(\lambda, y, \eta)|^2 dy$; since the operator mapping $v(x, y)$ into $e^{ix\cdot\lambda} v(x, y)$ is bounded from $H_{(0,-\frac{1}{2})}$ into itself with norm $\leq C|\lambda|^{\frac{1}{2}}$ we have

$$\|P\| \leq C \int |\lambda|^{\frac{1}{2}} \|P_\lambda\| d\lambda$$

$$\leq C \sup_\eta \int |\lambda|^{\frac{1}{2}} |\eta|^{-1} \int |b(\lambda, y, \eta)|^2 dy d\lambda$$

$$< \infty$$

and the proof of (3) is complete. $\square$

We now turn to the second part of the theorem (4) and (5) are reformulations on the well-known theorems of Giraud-Mikhlin and Calderon-Zygmund on singular integral operators; modern proofs can be found in E.M. Stein [80]. (Observe also that (4) of the Theorem could be derived from (1) as follows: if $a(x, \xi)$ is a symbol of order m which vanishes for ξ near 0, $e^{-y|\xi|}a(x,\xi)$ will be a pseudo-Poisson symbol of order m whose corresponding operator P when restricted to the hyperplane $y = 0$, yields precisely the pseudo-differential operator A of symbol $a(x, \xi)$; since P is bounded on Hölder spaces, so is A.) Thus it remains only to prove (6). We group the main steps in a simple lemma.

Lemma 1.53

(1) *There exists a pseudo-Poisson operator of order* 0 *and a constant* C *such that for all*
   $u \in B^P(\mathbb{R}^n) \cap \varepsilon'(K)$

$$C^{-1} \|u\|_{B^P(\mathbb{R}^n)} \leq \|Au\|_{L_1^P(\mathbb{R}_+^{n+1})} \leq C \|u\|_{B^P(\mathbb{R}^n)} \tag{1.95}$$

(2) *If* A *is a pseudo-differential operator of order* m *on* $\mathbb{R}^n$, *and* P *is a pseudo-Poisson operator of order* m', *then* $PA - (A \otimes 1)P$ *is a pseudo-Poisson operator of order* $m + m' - 1$.
(3) *Pseudo-Poisson operators of order* -1 *are bounded from* $L^P(\mathbb{R}^n)$ *to* $L_1^P(\mathbb{R}_+^{n+1})$.

Proof That the second inequality in (1.95) is valid for any pseudo-Poisson operator of order 0 is a consequence of (2) in Theorem 1.46.; now it follows from the basic existence theorem for the Dirichlet problem that there exists a function Au which is harmonic in $\mathbb{R}_+^{n+1}$, belong to $L_1^P(\mathbb{R}_+^{n+1})$, and admits u as a boundary value; in particular $\|u\|_{B^P} \leq$

$\|Au\|_{L_1^p(\mathbb{R}_+^{n+1})}$; since the map $u \to Au$ must be a pseudo-Poisson operator, the desired conclusion follows.

(2) is an easy consequence of the Kohn-Nirenberg formula for the composition of pseudo-differential operators. Indeed, it is easy to check that $PA - (A \otimes 1)P$ is a pseudo-Poisson operator with symbol given by

$$c(x, y, \xi) \sim \sum_{|\alpha| \geq 1} \frac{1}{\alpha!}(D_\xi^\alpha \sigma_p(x, y, \xi)\partial_x^\alpha \sigma_A(x, \xi) - D_\xi^\alpha \sigma_A(x, \xi)\partial_x^\alpha \sigma_p(x, y, \xi))$$

where the asymptotic convergence is uniform with respect to x and y.

(3) is trivial.

Now if A is a pseudo-differential operator of order 0 on $\mathbb{R}^n$ and if P is the pseudo-Poisson operator of (1) in Lemma 1.53., then

$$\|Au\|_{B^p} \leq C\|PAu\|_{L_1^p(\mathbb{R}_+^{n+1})}$$

$$\leq C(\|(A \otimes 1)Pu\|_{L^p(\mathbb{R}_+^{n+1})} + \|u\|_{L^p(\mathbb{R}^n)})$$

$$\leq C(\|Pu\|_{L^p(\mathbb{R}_+^{n+1})} + \|u\|_{L^p(\mathbb{R}^n)})$$

$$\leq C\|u\|_{B^p(\mathbb{R}^n)}$$

and the proof of Theorem 1.46. is complete. $\square$

We now study estimates that distinguish between different directions. Let a be a subvector bundle of $T(\mathbb{R}^n)$, and define a vector field X on $\mathbb{R}^n$ to be a-admissible if at any point $x \in \mathbb{R}^n$, $X(x) \in a_x$. Similarly, a vector field D on $\overline{\mathbb{R}_+^{n+1}}$ is a-admissible if $D(x, 0) \in a_x$ for any $x \in \mathbb{R}^n$. The notion of a-admissible curves on $\overline{\mathbb{R}_+^{n+1}}$ and $\mathbb{R}^n$ may be defined as in the Introduction; thus

$$S_{p,k}^\#(\mathbb{R}_+^{n+1}) = \{f \in L_1^p(\mathbb{R}_+^{n+1}) \; ; \; D_1^{\alpha_1} \cdots D_m^{\alpha_m} f \in L_1^p(\mathbb{R}_+^{n+1}), \; |\alpha| \leq k, \; D_j \; a\text{-admissible}\}$$

$$B_{p,k}^\#(\mathbb{R}^n) = \{f \in B^p \; ; \; X_1^{\alpha_1} \cdots X_m^{\alpha_m} f \in B^p(\mathbb{R}^n), \; |\alpha| \leq k, \; X_j \; a\text{-admissible}\}$$

For $0 < \alpha \leq 1$

$$\Gamma_\alpha^\#(\mathbb{R}_+^{n+1}) = \{f \in \Lambda_{\frac{\alpha}{2}}(\mathbb{R}_+^{n+1}) \; ; \; f \in \Lambda_\alpha[0, 1] \text{ when restricted to } a\text{-admissible curves}\}$$

$$\Gamma_\alpha^\#(\mathbb{R}^n) = \{f \in \Lambda_{\frac{\alpha}{2}}(\mathbb{R}^n); \; f \in \Lambda_\alpha[0, 1] \text{when restricted to} a\text{-admissible curves of the}$$
$$\text{boundary}\}$$

For $\alpha > 1$ the definitions are extended in the usual way. Obviously $\Gamma_\alpha^\#(\mathbb{R}^{n+1})$ is the space of restrictions of functions on $\Gamma_\alpha^\#(\mathbb{R}_+^{n+1})$ to the boundary while $B_{p,k}^\#$ in the same manner corresponds to $S_{p,k}^\#$.

Theorem 1.54

(1) *Any pseudo-Poisson operator of order* 0 *can be extended boundedly from* $B_{p,k}^{\#}(\mathbb{R}^n)$ *to* $S_{p,k}^{\#}(\mathbb{R}_+^{n+1})$ $(1 < p < \infty,\ k \geq 0)$.

(2) *Any pseudo-differential operator of order* 0 *is bounded from* $B_{p,k}^{\#}(\mathbb{R}^n)$ *to* $B_{p,k}^{\#}(\mathbb{R}^n)$ $(1 < p < \infty,\ k \geq 0)$.

(3) *If* P *is a pseudo-Poisson operator which is the parametrix of a coercive boundary value problem, then* P *is bounded from* $\Gamma_{\alpha+1}(\mathbb{R}^n) \cap \Lambda_{\alpha+\frac{1}{2}}(\mathbb{R}^n)$ *to* $\Gamma_{\alpha+1}(\mathbb{R}_+^{n+1}) \cap \Lambda_{\alpha+\frac{1}{2}}(\mathbb{R}_+^{n+1})$ *and from* $\Gamma_{\alpha+2}(\mathbb{R}^n) \cap \Lambda_{\alpha+1}(\mathbb{R}^n)$ *to* $\Gamma_{\alpha+2}(\mathbb{R}_+^{n+1}) \cap \Lambda_{\alpha+1}(\mathbb{R}_+^{n+1})$.

Proof Let D be an admissible vector field on $\overline{\mathbb{R}_+^{n+1}}$. D can be written as

$$D = X + b(x, y)Y$$

where X is an admissible vector field on $\mathbb{R}^n$, Y is some vector field on $\overline{\mathbb{R}_+^{n+1}}$, and $b(x, y) = 0$ for $y = 0$. Now

$$DPu(x, y) = XPu(x, y) + b(x, y)YPu(x, y)$$

In the right-hand side replace XP by $PX + [P, X]$, the commutator is bounded by $C\|u\|_{B^p}$ according to Lemma 1.53, while $\|PXu\|_{L_1^p(\mathbb{R}_+^{n+1})}$ can of course be bounded by $C\|Xu\|_{B^p(\mathbb{R}^n)}$. On the other hand

$$b(x, y)YPu(x, y) = \int b(x, y)Y(e^{ix\cdot\xi}a(x, y, \xi))\widehat{u}(\xi)\mathrm{d}\xi$$

defines a pseudo-Poisson operator of order 0; indeed, first assume that $Y = \frac{\partial}{\partial y}$. Since $D_x^{\alpha}b(x, y) = O(y)$ and $a(x, y, \xi)$ is a pseudo-Poisson symbol we have

$$|b(x, y)\frac{\partial a}{\partial y}(x, y, \xi)| \leq C|y\frac{\partial a}{\partial y}(x, y, \xi)| \leq C$$

and similarly for derivatives of higher order, which proves our assertion. If $Y = \frac{\partial}{\partial x_j}$ then the symbol to be considered is $b(x, y)\xi_j a(x, y, \xi)$ which also satisfies the desired estimates. Applying now Theorem 1.46, part (1) of this theorem follows immediately.

(2) can be proved as in (6) of Theorem 1.46. by using (1), and finally the proof of (3) can be found in P. Greiner and E.M. Stein [29] and D.-C. Chang, A. Nagel, E.M. Stein [13].

$\square$

If Ω is bounded domain in $\mathbb{R}^{n+1}$ with a smooth boundary, it is a simple matter to define $L_1^p(\Omega)$, $B^p(b\Omega)$, $\Lambda_{\alpha}(\Omega)$, etc. so as extend in a natural way the notions encountered in the case of $\mathbb{R}_+^{n+1}$, and similarly given a subbundle a of $T(b\Omega)$ non-isotropic Sobolev and Lipschitz spaces can be defined accordingly; Theorems 1.46. and 1.54. can then be reformulated in this context. The following particular cases are of special importance:

1. $a = T(b\Omega)$; the spaces $S_{p,k}^{\#}$ and $B_{p,k}^{\#}$ then reduce to well-known spaces
   $$L_{k+1}^p(\Omega) = \{f \in L^p(\Omega); \, p(x, D)f \in L^p(\Omega) \text{ for any differential operator } p(x, D) \text{ of order } \leq k+1\}$$
   $$B_k^p(\Omega) = \{f \in B^p ; \, p(x, D)f \in B^p(\Omega^p(\Omega)) \text{ for any differential operator } p(x, D) \text{ of order } \leq k\}.$$
2. $\Omega \subset \mathbb{C}^n (n \geq 2)$, and $a = T^{1,0}(b\Omega) \oplus T^{0,1}(b\Omega)$; $S_{p,k}^{\#}$, $\Gamma_\alpha^{\#}(\Omega)$ and $\Gamma_\alpha^{\#}(b\Omega)$ are then the non-isotropic Sobolev and Hölder spaces of the Introduction.

Operators of Fuchs Type

2

2.1 Degenerate Operators and Singular Manifolds

In the following Sect. 2.1 we will outline some intiutive ideas of the analysis of pseufo-differential operators (PDOs) on singular manifolds, before we pass to technicalities in special cases and more references. PDOs on smooth manifolds in local coordinates are known to be dominated by principal or complete symbols. On the other hand, configurations with singularities (briefly called singular manifolds) give rise to various types of degenerate operators, the symbols of which do not belong to the standard classes. An example of conical singularities are the corners of the boundary M of a triangle in $\mathbb{R}^2$. Those are denoted by M_{sing}. More generally a set M, endowed with a metric g, is a manifold with singularities $M_{\mathrm{sing}} \subset M$, when $M_{\mathrm{reg}} := M \setminus M_{\mathrm{sing}}$ is (smooth) Riemannian with respect to the restriction g_{reg} of g to M_{reg}, and M_{sing} is a "thin" subset. This set may consist of finitely many so-called conical points, or of some higher-dimensional manifold (e.g., a boundary or an edge). Moreover, the metric g may be non-smooth close to M_{sing}. This is, for the moment, a vage definition, which needs more precise conditions on the behavior of g near M_{sing}. Those will be considered later on in the context of stretched manifolds. Further examples of manifolds with conical singularities are graphs G, where G_{sing} is a finite set of points, and G_{reg} is of dimension 1. Extra conditions will describe, on how G_{sing} is embedded in G. In particular, we rule out cuspidal singularities. A regular example $G \subset \mathbb{R}^3$ is defined as the union of all one-dimensional edges of the surface of the unit cube in $\mathbb{R}^3$. In this case G_{sing} consists of the corner points of G. Note in particular, that the edges have common corner points in transversal position. Analogous properties hold for other systems of one-dimensional edges belonging to more general polyhedra in $\mathbb{R}^N$, $N \in \mathbb{N}$. More examples are generated by assuming $G \subset \Omega$ for an open neigborhood $\Omega \subset \mathbb{R}^N$. Let $\delta : \Omega \to \tilde{\Omega}$ be a diffeomorphism between open subsets of $\mathbb{R}^N$; then $\tilde{G} := \delta(G)$ is such an example.

© Springer Nature Switzerland AG 2025
D.-C. Chang, B.-W. Schulze, *Analysis on Manifolds with Singularities*,
Pseudo-Differential Operators 16, https://doi.org/10.1007/978-3-032-01890-8_2

Pseudo-differential operators (PDOs) P on a singular manifold M are basically given on M_{reg} as ordinary PDOs, for the moment denoted by P_{reg}, are considered together with some extension to M_{sing}. Such an extension may be very individual; the meaning depends on how the operator P_{reg} induces an action between spaces of distributions over M_{reg}, combined with a certain control close to M_{sing}. Such a procedure depends both on the symbol of P_{reg} and on the nature of the above mentioned distribution spaces close to M_{sing}. This is a relatively complicated aspect. However, (in order to illustrate a classical special case), if M_{reg} is the open interior of a compact smooth manifold M with smooth boundary ∂M, identified with M_{sing}, in the case of boundary problems for a PDO of integer order $\mu \leq 0$ with the transmission property at the boundary (see more details below), then the operator $P : H^s(M) \to H^{s-\mu}(M)$ between standard Sobolev spaces over M acts via the truncation quantization, cf. [9] and other research on this topic, e.g., the work of G. Grubb [30] and a recent joint paper with J. Seiler [72]. On the other hand, in the present case of conical singularities an example is $M := \overline{\mathbb{R}}_+$ with $M_{\mathrm{reg}} := \mathbb{R}_+$ and $M_{\mathrm{sing}} := \{0\}$. A differential operator P of Fuchs type on M is determined by a "degenerate" symbol of the form $p := p(r, r\rho)$ in the variable $r \in M$ and the co-variable $\rho \in \mathbb{R}$, such that $r\rho$ represents the action $-ir\frac{\partial}{\partial r}$, which is also called "totally characteristic derivative" by some authors. It is typical for symbols of operators close to singularities of manifolds M, that covariables may be multiplied with variables on the manifold itself, as is the case of operators of Fuchs type. This is for the moment a crude description; more details in edge and corner cases will be given later on. Interpreting the above-mentioned totally characteristic operator P as $P_{\mathrm{reg}} : C^\infty(\mathbb{R}_+) \to C^\infty(\mathbb{R}_+)$ for the case $\mathbb{R}_+ := M_{\mathrm{reg}}$, then we have another example of the above-mentioned operator convention, compared with the truncation quantization. Below we shall give an alternative description in more general context of other singular manifolds and adapted notation, but the scheme, that was just described, will be valid also on manifolds with edges. This will be an analogue of boundary problems, though without the transmission property. The following material can be seen as a special edge theory for edges of dimension 0, corresponding to $r = 0$. On the other hand, in slightly modified form, it plays a role as an edge symbolic calculus, analogously to so-called boundary symbols in Boutet de Monvel's calculus of pseudo-differential boundary problems with the transmission property at the boundary.

Analysis on manifolds M with singularities (pseudo-manifolds in alternative notation) in the case of conical points will be developed in localized form in a neighborhood U of a chosen conical singularity, which is so small, that it makes sense to talk about the distance $r > 0$ of a point in U. Local coordinates are then chosen in such a way, that r varies over the whole half-axis $\mathbb{R}_+$. This is for the moment an intuitive description. Later one we come back to an identification of U in stretched form like $\mathbb{R}_+ \times X \ni (r, x)$, where X is a closed compact smooth manifold, and the latter Cartesian product is regarded as a subset of $\overline{U} := \overline{\mathbb{R}}_+ \times X$. In addition, our calculus on U will be organized in terms of operator-valued Mellin symbols. Those will correspond to the above-mentioned degenerate symbols of operators in dimension 1, and we will consider the pseudo-differential case. For simplicity we assume, that M only has one conical singularity. Then another step for studying operators globally on M, may use an open covering of M, containing U together with an open set of M_{reg} and a subordinate partition of unity, together with standard pseudo-differential operators on M_{reg}. Then we can form operators globally on M as usual, by considering sums with terms referring to U and contributions from the regular part of M. Here we focus on the contribution from U close to $r = 0$, but we also observe, what happens for $r \to +\infty$ in these special local coordinates of U. We also pass to holomorphic extensions in the Mellin covariable z, representing $-i r \frac{\partial}{\partial r}$ off some set $V \subset \mathbb{C}$, supporting an "asymptotic quantity". We refer to ideas, suggested by the material of [61] in the framework of analytic functionals, here in a simpler distributional context. Furthermore, we formulate symbols, obtained by a kernel cut-off operator, which produces holomorphic dependence in the co-variable $z \in \mathbb{C}$, modulo an error term of order $-\infty$.

2.2   Kernel Cut-off

We first fix some notation and technicalities on the Mellin transform, including the symbolic structure of Mellin pseudo-differential operators. The latter notation means, that the Fourier transform F, involved in pseudo-differential operators (PDOs) in $\mathbb{R} \ni x$ with co-variable $\xi \in \mathbb{R}$ is replaced by the Mellin transform

$$(M_\beta u)(z) := \int_0^\infty r^z u(r) \frac{dr}{r}, \tag{2.1}$$

acting, for instance, as a continuous map $M_\beta : C_0^\infty(\mathbb{R}_+) \to \mathcal{A}(\mathbb{C})$ between the involved functions in $C_0^\infty(\mathbb{R}_+)$, $r \in \mathbb{R}_+$, and holomorphic functions $\mathcal{A}(\mathbb{C})$ in $z \in \mathbb{C}$, running on

$$\Gamma_{\frac{1}{2}-\beta} := \{z \in \mathbb{C} : \text{Re}(z) = \frac{1}{2} - \beta\} \tag{2.2}$$

for any fixed (so-called) weight $\beta \in \mathbb{R}$. Here (2.2) is called a weight-line and (2.1) the corresponding weighted Mellin transform. For $\beta = 0$ we also write M instead of M_0. Concerning material on M between spaces of distributions on the real axis we also refer to F. Jeanqartier [35], or to papers by numerous authors, e.g., of V.A. Kondratyev [40], B.A. Plamenevskij [53], J. Cheeger [17], G.I. Eskin [24], R.B. Melrose and G.A. Mendoza [50], R. Mazzeo [46], Ju.V. Egorov and B.-W. Schulze [23], Ch. Dorschfeldt [22],I. Witt [83], J. Gil, B.-W. Schulze and J. Seiler [27, 28], T. Krainer [41], S. Khalil [37], D.-C. Chang, S. Khalil and B.-W. Schulze [15, 16], B.-W. Schulze and J. Seiler [72], and many others. In particular, it is well-known, that M_β induces an isomorphism

$$M_\beta : L^{2,\beta}(\mathbb{R}_+) \to L^2(\Gamma_{\frac{1}{2}-\beta}), \tag{2.3}$$

for $L^{2,\beta}(\mathbb{R}_+) := r^\beta L^2(\mathbb{R}_+)$, with $L^2(\mathbb{R}_+)$ being the space of square integrable functions on $\mathbb{R}_+$. The latter will locally represent a cone-axis close to some conical singularity, see details below. Note also that for $u \in C_0^\infty(\mathbb{R}_+)$ we have the relation $(M_\beta u)(z) = M_0(r^{-\beta}u)(z + \beta)$. Sometimes, for convenience, we consider the case $\beta = 0$ and indicate by $(M_{r \to z}u)(z) = f(z)$ the transfer of the variable r in the involved $u(r)$ and the co-variable $z \in \Gamma_{\frac{1}{2}-\beta}$ in $f(z)$ by subscript $r \to z$. For later reference we set

$$(S_\beta u)(t) = e^{-(1/2-\beta)t}u(e^{-t})$$

and

$$(\Phi_\beta v)(1/2 - \beta + i\rho) := v(\rho).$$

We define $\mathcal{S}(\Gamma_{\frac{1}{2}-\beta}) = \Phi_\beta(\mathcal{S}(\mathbb{R}))$, with $\mathcal{S}(\mathbb{R})$ being the Schwartz space on the real axis. This implies the relation

$$(M_\beta u)(1/2 - \beta + i\rho) = (\Phi_\beta \circ F \circ S_\beta u)(1/2 - \beta + i\rho)$$

for the Fourier transform F on the real t-axis, and we see, in particular, for $\mathcal{T}_\beta(\mathbb{R}_+) := S_\beta^{-1}(\mathcal{S}(\mathbb{R}))$, that

$$M_\beta : \mathcal{T}_\beta(\mathbb{R}_+) \to \mathcal{S}(\Gamma_{\frac{1}{2}-\beta})$$

induces an isomorphism between the involved Fréchet spaces. We then have

$$(M_\beta^{-1}f)(r) = \frac{1}{2\pi i}\int_{\Gamma_{\frac{1}{2}-\beta}} r^{-z}f(z)dz. \tag{2.4}$$

Next we formulate Mellin symbols and associated Mellin PDOs. In particular, we have holomorphic Mellin symbols in $z \in \mathbb{C}$ and the subsequent kernel cut-off construction as a preparation for symbols with asymptotic data. General symbols, not only for the $\mathbb{R}_+$-case, but also including variables $x \in \mathbb{R}^n$ referring to a future link X of a conical singularity, playing the role of local coordinates x on X and the Fourier transform $F = F_{x \to \xi}$. The space

$$S_B^\mu(\overline{\mathbb{R}}_+ \times \mathbb{R}^n \times \Gamma_{\frac{n+1}{2}-\beta} \times \mathbb{R}^n)$$

in variables $(r, x) \in \mathbb{R}_+ \times \mathbb{R}^n$ and covariables $(\rho, \xi) \in \mathbb{R} \times \mathbb{R}^n$ is defined as

$$C_B^\infty(\overline{\mathbb{R}}_+, E) := C_B^\infty(\mathbb{R}_+, E) \cap C_b^\infty(\mathbb{R}_+, E), \tag{2.5}$$

for some Fréchet space E, here for the space of classical symbols

$$E := S_{\mathrm{cl}}^{\mu}(\mathbb{R}_x^n \times \mathbb{R}_\rho \times \mathbb{R}_\xi^n),$$

$\rho := \mathrm{Im}(z)$ for $z \in \Gamma_{\frac{n+1}{2}-\beta}$. Moreover, $C_{\mathrm{B}}^{\infty}(\mathbb{R}_+ \ldots)$ indicates functions $u(r, \ldots)$ in $r \in \mathbb{R}_+$, which are bounded, including all derivatives $(r\partial_r)^k u(r, \ldots)$. Those symbols h (first, for $\beta = 0$) induce analogues of PDOs

$$\mathrm{op}_{(r,x)}(h) : u(r, x) \to M^{-1} F^{-1}\big(h(r, x, z, \xi) M_{r' \to z} F_{x' \to \xi'} u(r', x')\big). \tag{2.6}$$

Here M^{-1} is combined with the pseudo-differential action in (x, ξ). In more concise interpretation we will employ Mellin symbols, taking values in Fréchet spaces, or in associated PDOs, where we mainly point out the Mellin action. This notation will be modified in different applications below. Relation (2.6) will refer to $u(r, x) \in \mathcal{H}^s(\mathbb{R}_+ \times \mathbb{R}^n) =: \mathcal{H}^s(\mathbb{R}_+^{1+n})$, which means the completion of $C_0^{\infty}(\mathbb{R}_+^{1+n})$ with respect to the norm

$$\|u\|_{\mathcal{H}^s(\mathbb{R}_+^{1+n})} := \Big\{ \int_{\Gamma_{\frac{n+1}{2}}} \int_{\mathbb{R}^n} (1 + |z|^2 + |\xi|^2)^s |MFu(z, \xi)|^2 dz d\xi \Big\}^{1/2}.$$

This concerns the case $\beta = 0$, and $h(r, x, z, \xi)$ is said to be a left symbol in the corresponding Mellin/Fourier oscillatory integral, also indicated by h_{L}. Let us set

$$\mathcal{H}^{s,\beta}(\mathbb{R}_+^{1+n}) := r^\beta \mathcal{H}^s(\mathbb{R}_+^{1+n})$$

for arbitrary reals s, β. Links of conical singularities will belong to category of oriented compact smooth manifolds X, denoted by $\mathfrak{M}_0$. For any $X \in \mathfrak{M}_0$ of dimension n we define the weighted space $\mathcal{H}^{s,\beta}(\mathbb{R}_+ \times X)$ to be the space of all distributions on $\mathbb{R}_+ \times X$ restricting to $\mathcal{H}^{s,\beta}(\mathbb{R}_+ \times \mathbb{R}^n)$ for any coordinate neighborhood U of X, where $x \in \mathbb{R}^n$ plays the role of local coordinates on U. The definition may be given in terms of a finite open covering of X and a subordinate partition of unity. Note that

$$\mathcal{H}^{s,\beta}(\mathbb{R}_+ \times X) \subset H_{\mathrm{loc}}^s(\mathbb{R}_+ \times X).$$

Below we admit vector-valued Mellin symbols $h(r, z, x, \xi, \lambda)$ with parameter $\lambda \in \mathbb{R}^d$ of the class

$$C_B^{\infty}(\overline{\mathbb{R}}_+, L_{\mathrm{cl}}^{\mu}(X; \Gamma_{\frac{n+1}{2}-\beta} \times \mathbb{R}^d; E, \tilde{E})),$$

with $r \in \overline{\mathbb{R}}_+, z \in \Gamma_{\frac{n+1}{2}-\beta}, \lambda \in \mathbb{R}^d$, for some $X \in \mathfrak{M}_0$ of dimension n. We will also write

$$\mathrm{op}_M^\beta(h)(\lambda)$$

for the corresponding Mellin action. Locally on $\mathbb{R}_+ \times \mathbb{R}^n$ a pseudo-differential action (first without parameters and for double symbols $h_{\mathrm{D}}(r, r', z)$) is given as a Mellin oscillatory integral

$$\mathrm{op}_M^\beta(h_{\mathrm{D}})u(r) := \int \int_0^\infty \left(\frac{r'}{r}\right)^{-(\frac{n+1}{2} - \beta + i\rho)} h_{\mathrm{D}}\left(r, r', \frac{1}{2} - \beta + i\rho\right) u(r') \frac{dr'}{r'}\, đ\rho, \qquad (2.7)$$

for $đ\rho := \frac{1}{2\pi i}d\rho$ (variables x, ξ are suppressed in this notation.) Operators (2.7) induce continuous maps

$$\mathrm{op}_M^\beta(h_{\mathrm{D}}) : \mathcal{H}^{s,\beta}(\mathbb{R}_+^{1+n}) \to \mathcal{H}^{s-\mu,\beta}(\mathbb{R}_+^{1+n}). \qquad (2.8)$$

Below we frequently use operator-valued left Mellin symbols $h(r, z) = h_{\mathrm{L}}(r, z)$, acting in spaces $\mathcal{H}^s(X^\wedge)$ for compact $X \in \mathfrak{M}_0$, where the (r, z)-wise action is pseudo-differential and locally takes place first in the variables $x \in \mathbb{R}^n$ which gives us a map $H^s(\mathbb{R}^n) \to H^{s-\mu}(\mathbb{R}^n)$, and then via a partition of unity, referring to charts, belonging to an open covering of X by coordinate neighborhoods, combined with a subsequent summation, which gives rise to a family of operators $h(r, z) : H^s(X) \to H^{s-\mu}(X)$. Assuming an extra dependence of h on $(y, r\eta)$ for $(y, \eta) \in \Omega \times \mathbb{R}^q$ for open $\Omega \subseteq \mathbb{R}^q$ and multiplying everything by $\omega(r|\eta|)$ we obtain actions

$$\omega(r|\eta|)\mathrm{op}_M(h(0, y, z, r\eta)) : \mathcal{H}^s(X^\wedge) \to \mathcal{H}^{s-\mu}(X^\wedge).$$

This will be applied below in the edge calculus, where Mellin symbols h admit a restriction to $r = 0$ and, if necessary, and an extra multiplicative factor $r^{-\mu}$. Another aspect will be, that the operator families are considered together with an action of some one-parameter group $\{\kappa_\iota\}_{\iota \in \mathbb{R}_+}$ of strongly continuous operators on some Hilbert space H, here $H = \mathcal{H}^s(X^\wedge) \ni u(r, x)$ with $(\kappa_\iota u)(r, x) = \iota^{(n+1)/2}u(\iota r, x)$. In general, the group action is assumed to be strongly continuous and to satisfy the relations $\kappa_\iota \kappa_\vartheta = \kappa_{\iota\vartheta}$ for all $\iota, \vartheta \in \mathbb{R}_+$ and $\kappa_1 = \mathrm{id}_H$. There is then the aspect of operator-valued symbols of the class $S^\mu(\Omega \times \mathbb{R}^q; H, H') \ni a(y, \eta)$, for Hilbert spaces H and H' with group actions $\{\kappa_\iota\}_{\iota \in \mathbb{R}_+}$ and $\{\kappa_\iota'\}_{\iota \in \mathbb{R}_+}$, respectively, defined by symbolic estimates

$$\|(\kappa')^{-1}(\eta)\{D_y^\alpha D_\eta^\beta a(y, \eta)\}\kappa(\eta)\|_{\mathcal{L}(H,H')} \le c[\eta]^{\mu-\beta},$$

where $\kappa(\eta) := \kappa_{[\eta]}$ and $\eta \to [\eta]$ is some non-vanishing function with $[\eta] = |\eta|$ for $|\eta| \ge 1$.

2.3   Symbols with Distributional Asymptotic Densities

Now we look at specific operator-valued symbols

$$h(r, z) \in C_B^\infty(\overline{\mathbb{R}}_+, L_{cl}^\mu(X; \Gamma_{\frac{n+1}{2}-\beta}))$$

for some closed compact manifold X of dimension n, which extend from $z \in \Gamma_{\frac{n+1}{2}-\beta}$ to a wider region of the complex plane $\mathbb{C}$, for instance as holomorphic or meromorphic functions. In this connection we first consider the holomorphic case and so-called kernel cut-off operators, for convenience, for $\beta = 0$.

For a Mellin PDO

$$C_0^\infty(\mathbb{R}_+ \times X) \ni u \to \operatorname{op}_M^{(n+1)/2}(h)u = \iint \Big(\frac{r}{r'}\Big)^{(n+1)/2-i\rho} h_D(r, r', z)u(r')\frac{dr'}{r'}d\!\!\!{}^-\!\rho,$$

with double symbol $h_D(r, r', z)$, which are not necessarily holomorphic in z, the kernel

$$k(r, r') = \int \Big(\frac{r}{r'}\Big)^{(n+1)/2-i\rho} h_D(r, r', z)d\!\!\!{}^-\!\rho$$

has its singular support on the diagonal $r = r'$. We also write

$$k(r, r') = k(r, r', \zeta)|_{\zeta=r/r'}$$

with

$$k(r, r', \zeta) := \int \zeta^z h_D(r, r', z)d\!\!\!{}^-\!\rho.$$

From the kernel we can return to the amplitude function by forming

$$h_D(r, r', z) := M_{\zeta \to z}k(r, r', \zeta).$$

Proposition 2.1 *Let $\varphi(r/r')$ be a function, such that $\varphi(\zeta) \in C_0^\infty(\mathbb{R}_{+,\zeta})$ and $\varphi(\zeta) \equiv 1$ in a neighborhood of $\zeta = 1$. Then we have*

$$h_O(r, r', z) := (M_{\zeta \to z}(\varphi(\frac{r}{r'})k(r, r', \zeta)) \in C_B^\infty(\overline{\mathbb{R}}_+ \times \overline{\mathbb{R}}_+, M_O^\mu(X)).$$

and

$$h_{-\infty}(r, r', z) := (M_{\zeta \to z}(1 - \varphi(\frac{r}{r'}))k(r, r', \zeta)) \in C_B^\infty(\overline{\mathbb{R}}_+ \times \overline{\mathbb{R}}_+, L^{-\infty}(X; \Gamma_{\frac{n+1}{2}})).$$

This concerns $h \in C_B^\infty(\overline{\mathbb{R}}_+ \times \overline{\mathbb{R}}_+, L_{cl}^\mu(X; \Gamma_{\frac{n+1}{2}}))$. In particular, we obtain

Theorem 2.2 *For every $h \in C_B^\infty(\overline{\mathbb{R}}_+ \times \overline{\mathbb{R}}_+, L_{cl}^\mu(X; \Gamma_{\frac{n+1}{2}}))$ there is a $h_O \in C_B^\infty(\overline{\mathbb{R}}_+ \times \overline{\mathbb{R}}_+, M_O^\mu(X))$ such that*

$$\mathrm{op}_M^{(n+1)/2}(h) = \mathrm{op}_M^{(n+1)/2}(h_O)$$

modulo a smoothing operator with symbol $h_{-\infty}(r, r', z)$.

Another way of organizing relations of this type is to replace the function $\varphi(r/r')$ by $\varphi(\zeta)$. In particular, as a special case, we can consider symbols in $M_O^\mu(X)$.

Proposition 2.3 *Proposition 2.1 reduces to a map*

$$H(\varphi) : L_{cl}^\mu(X; \Gamma_{\frac{n+1}{2}}) \to M_O^\mu(X), \tag{2.9}$$

called a kernel cut-off operator.

A similar notation will be used for the corresponding map in Proposition 2.1.

Remark 2.4 We may also consider symbols, taking values in $L_{cl}^\mu(X; \Gamma_{\frac{n+1}{2}} \times \mathbb{R}^d)$, i.e., with parameters $\lambda \in \mathbb{R}^d$.

In particular, we write $M_O^\mu(\mathbb{R}^q) := H(\varphi)L_{cl}^\mu(X; \Gamma_0 \times \mathbb{R}^q)$ for $\dim(X) = 0$. In addition, the Mellin action can refer to any weight $\beta \in \mathbb{R}$. Then we use the notation $\mathrm{op}_M^{(n+1)/2-\beta}(h)(\lambda)$ rather than $\mathrm{op}_M^{(n+1)/2}(h)$ for $n := \dim X$. If we say nothing else, for brevity we also omit indicating the parameter. Observe that for $(T^\delta h)(t, t', z) = h(t, t', z + \delta)$ we have

$$\mathrm{op}_M^\beta(h)r^{-\delta} = r^{-\delta}\mathrm{op}_M^{\beta+\delta}(T^\delta h)$$

and for any β, γ

$$\mathrm{op}_M^\beta(h) = \mathrm{op}_M^\gamma(h)$$

on $C_0^\infty(X^\wedge)$. Moreover, observe that

$$M_O^{-\infty}(X; \mathbb{R}^q) = C^\infty(X \times X)\hat{\otimes}_\pi M_O^{-\infty}(\mathbb{R}^q).$$

Motivated by degenerate operators, where symbols in an axial variable $r \in \mathbb{R}_+$ and the corresponding co-variable $\rho = \mathrm{Im}(z)$ for $z \in \Gamma_{\frac{n+1}{2}-\beta}$ depend on $r\rho$ we now consider

Mellin operators with symbols in $M_O^\mu(X)$, first for the case $n = \dim(X) = 0$. In elliptic theory there will be generated also meromorphic Mellin symbols and other generalizations. Those are connected with terms of the form $\omega(r)r^{-p}\log^k(r)$ for certain $p \in \mathbb{C}, k \in \mathbb{N}$, belonging to so-called discrete asymptotics of solutions. Writing

$$\omega(r) = \varepsilon(r) + (\omega(r) - \varepsilon(r))$$

with the characteristic function $\varepsilon(r)$ of the interval $[0.1]$ on the real axis, then we obtain the relation

$$M\{\omega(r)r^{-p}\log^k(r)\}(z) = \frac{(-1)^k k!}{(z-p)^{k+1}} + a_k(z-p)$$

for some $a_k(z) \in \mathcal{A}(\mathbb{C})$, with M being the weighted Mellin transform for some suitable weight. Let $f(z) := M\{\omega(r)r^{-p}\log^k(r)\}(z)$ and let C be a smooth clockwise oriented curve surrounding $\{p\}$. Then for every $h \in \mathcal{A}$ we have

$$\langle \zeta, h \rangle = \frac{1}{2\pi i}\int_C f(z)h(z)dz = (-1)^k \frac{d^k}{dz^k}h(z)|_{z=p}\,,$$

which is just (up to $(-1)^k$) equal to the kth derivative of the Dirac distribution at $z = p$. Inserting in the latter relation the function $h(z) = r^{-z}$ we get

$$\langle \zeta, r^{-z} \rangle = r^{-p}\log^k(r)\,.$$

Thus we obtain altogether

Proposition 2.5 *Let $f(z)$ be a meromorphic function with poles p_j of multiplicity m_j+1, $j = 0,\ldots,N$, and let C be a curve of similar kind as before, clockwise surrounding $K = \{p_0\} \cup \{p_1\} \cup \ldots \cup \{p_N\}$. Moreover, let $(-1)^k k! c_{jk}$ be the Laurent coefficients of f at $(z-p_j)^{-(k+1)}, 0 \le k \le m_j$. Then formula $\langle \zeta, h \rangle = \frac{1}{2\pi i}\int_C f(z)h(z)dz$ for any $h \in \mathcal{A}$ represents an element $\zeta \in \mathcal{E}'(U)$ for any open set U containing the set K such that $\mathrm{supp}(\zeta) = K$, and we have*

$$\langle \zeta, h \rangle = \sum_{j=0}^{N}\sum_{k=0}^{m_j}(-1)^k c_{jk}\frac{d^k}{dz^k}h(z)|_{z=p_j}\,, \tag{2.10}$$

for any holomorphic function h, in particular,

$$\langle \zeta, r^{-z} \rangle = \sum_{j=0}^{N}\sum_{k=0}^{m_j} c_{jk}r^{-p_j}\log^k(r)\,. \tag{2.11}$$

Similar asymptotics may be observed in connection with meromorphic Mellin symbols, appearing in parametrices of elliptic (in simplest cases, Fuchs type) differential operators.

Example 4 As an example we consider a family of parameter-dependent differential operators

$$r^{-\mu} \sum_{j=0}^{\mu} a_j(y)\left(-r\frac{\partial}{\partial r}\right)^j u(r, y) = m(r, y). \tag{2.12}$$

on the half-axis $\mathbb{R}_+$, where y is regarded as a parameter. We compute the asymptotics for $r \to 0$.

Assume $a_j(y) \in C^\infty(\Omega)$, $\Omega \subseteq \mathbb{R}^q$. Let $a(y, z) := \sum_{j=0}^{\mu} a_j(y)z^j \neq 0$ for all $y \in \Omega$, $z \in \Gamma_{\frac{1}{2}}$. Set $m(r, y) = \omega(r)r^{-p(y)}\log^k(r)$ for some $p(y) \in C^\infty(\Omega)$ and $k \in \mathbb{N}$. Let $\mathrm{Re}(p(y)) < \frac{1}{2}$ for all $y \in \Omega$. Equation (2.12) can be solved in the form

$$u(r, y) = M^{-1}a^{-1}(z, y)M(r^\mu m)(z, y), \tag{2.13}$$

with the Mellin transform $M : L^2(\mathbb{R}_+) \to L^2(\Gamma_{\frac{1}{2}})$. For any fixed y we find the asymptotics of $u(r, y)$ by calculating the Laurent expansion of $f(z, y) := a^{-1}(z, y)M(r^\mu m)(z, y)$ in z. First note, that $M(r^\mu m)(z, y)$ is meromorphic and has a pole at $p(y) - \mu$ of multiplicity 1. Also $f(z, y)$ is meromorphic, where the zeros of $a(z, y)$ in z contribute to a family of y dependent poles $p_1(y), \ldots, p_{\mu+1}(y)$ of $f(z, y)$. They are in general not smooth in y, and may be of branching behavior with varying y. We may have jumping multiplicities of the poles of $f(z, y)$ with varying y with a corresponding irregular dependence on y. Assuming for convenience that those poles vary in a compact set K where C is the boundary of some open set U, and if we note that C is a compact curve, clockwise surrounding K, then the expression

$$\langle \zeta, h \rangle = \frac{1}{2\pi i} \int_C f(z, y)h(z)dz \tag{2.14}$$

represents a y-dependent family $\zeta(y)$ of elements in $\mathcal{E}'(U)$, for the above-mentioned U containing the set K, which are y-wise derivatives of finite order of the Dirac distribution at points in K, multiplied by some coefficients. This gives us altogether the y-wise discrete and possibly jumping asymptotics of the solution (2.13) as $r \to 0$. When we ignore the dependence of y (i.e., consider the case of constant dependence on y), then $K = \{p\}$ equals a single point, and a simple example is, that the Dirac distribution is concentrated on p. On the other hand, we find a distribution supported by the compact set C, satisfying relation (2.14) with a density $v \in C^\infty(C)$. This phenomenon is similar to the sweeping out principle (balayage) in classical potential theory, referring to a fundamental solution of the Laplacian in $\mathbb{R}^n$. It says, roughly speaking, that when we consider a smooth domain $U \subset \mathbb{R}^n$ with boundary C and a compact subset $K \subset \overline{U}$, for every measure μ supported by K there is a measure v supported by C such that the Newtonian potential of μ coincides with that of v in $\mathbb{R}^n \setminus \overline{U}$. In particular, for $K = \{p\}$ and $\mu := \delta_p$, the measure v is just the harmonic measure belonging to the point $\{p\}$. The analogue to the present situation is, that here we speak about z^{-1}, a fundamental solution of the Cauchy-Riemann operator, rather than about the Newtonian kernel as a fundamental solution of the Laplace operator.

Let us finally illustrate variable and branching asympotics by some example of a family of meromorphic functions with variable and branching poles in the complex plane. The example has the following structure. Choose three continuous functions $p_j(t)$, $j = 1, 2, 3$ with values in $\mathbb{C}$, where $t \in \mathbb{R}$. Then $u(z) = \frac{z - p_1(t)}{(z - p_2(t))(z - p_3(t))}$ is a meromorphic family with poles at $z = p_2(t)$ and $z = p_3(t)$ provided that $p_1(t) \neq p_2(t)$ and $p_1(t) \neq p_3(t)$. Otherwise, e.g., when $p_1(t) = p_2(t)$ the corresponding pole at $z = p_3(t)$ is simple. We see that the flow of poles depending on t is variable and branching, according to the position of $p_j(t)$, $j = 1, 2, 3$ in the complex plane.

2.4   Conical Singularities in Stretched Description

Our next objective is to give an idea of manifolds with conical singularities. First let $\mathfrak{M}_0$ be a category of smooth compact oriented manifolds. Elements of $\mathfrak{M}_0$ will be regarded as links of conical singularities as follows. First we use the fact, that any element $X \in \mathfrak{M}_0$ of dimension n can be embedded as a smooth submanifold of the unit sphere $\mathbb{S}^N$ of $\mathbb{R}^{N+1}$ for sufficiently large N. For convenience we identify X with its image in $\mathbb{S}^N$. We now form the union of all infinite straight half-lines, containing points of $X \subset \mathbb{S}^N$ and the origin $\{0\} \in \mathbb{R}^{N+1}$. Let us denote the arizing infinite cone by $X^\triangle$. This is a C^∞-manifold of dimension $n + 1$ for $n = \dim(X)$ outside the conical point $\{0\} \in \mathbb{R}^{N+1}$. For purposes below we introduce the finite subcones

$$X_h^\triangle := \{\tilde{x} \in X^\triangle : |\tilde{x}| < h\}$$

for any $h > 0$. Now a compact (pseudo)-manifold M with conical singularity $\{0\}$ and link X is defined by a gluing together (in an obvious smooth process) a smooth manifold B of dimension $n + 1$ with boundary $\partial B = X$ and $X_h^\triangle$ for some fixed $h > 0$ along $\{h\} \times X$. In other words, the boundary X of B is identified with $\{\tilde{x} \in X^\triangle : |\tilde{x}| = h\}$ for some $h > 0$ in such a way, such that we obtain altogether a smooth manifold M (except for the exceptional point). The corresponding manifold with conical point is (roughly speaking) formed by B and $X_h^\triangle$ by identifying both spaces along the common sub-manifold X. In a similar manner we can form the associated stretched manifold $\mathbb{M}$ by gluing together B with the cylinder

$$\mathbb{X}_h^\Delta := [0 \times h] \times X$$

along $X \subset B$ and $\{h\} \times X$, respectively, in a smooth way.

In other words, the conical singularity in stretched form is replaced by a cylinder with cross section X. Note that X may have several connected components. On the stretched manifold $\mathbb{M}$ (identified with a corresponding manifold M with conical singularities) we now establish a variant of the so-called cone-algebra, consisting of pseudo-differential operators, containing all differential operators of Fuchs type, motivated by Laplace-Beltrami operators belonging to the conical geometry of the underlying M, together with their pseudo-differential paramatrices and the generated operator algebra.

2.5   The Cone Calculus

Analysis close to the conical point is essentially formulated in terms of

$$(r, x) \in [0 \times h] \times X$$

by using the Mellin transform along $(0 \times h]$ or $\mathbb{R}_+$, while along X we employ the Fourier transform and associated pseudo-differential formulations. We focus on elliptic theory, starting with differential operators. Typical examples are so-called operators of Fuchs type, in particular Laplace-Beltrami operators close to conical singularities. Those are of the form (for $\mu = 2$)

$$A = r^{-\mu} \sum_{k=0}^{\mu} a_k(r)(r\frac{\partial}{\partial r})^k \tag{2.15}$$

with coefficients $a_k(r) \in C^\infty(\overline{\mathbb{R}}_+, \text{Diff}^{\mu-k}(X))$ with $\text{Diff}^j(X)$ being the space of all differential operators of order j on the compact link $X \in \mathfrak{M}_0$ in its canonical Fréchet topology. When we perform elliptic theory in this context and consider parametrices, we speak on degenerate operators in the sense, that the covariable belonging to $r \in \mathbb{R}_+$ corresponds to $z \in \mathbb{C}$, varying on $\Gamma_{\frac{1}{2}+\beta}$ or $\Gamma_{\frac{n+1}{2}+\beta}$, for some real β, according to $n = \dim(X)$. Fuchs type operators like (2.15) and their Fredholmness in weighted Sobolev spaces on $\mathbb{M}$ are well-known, and we refer here to some basic results, to be used below for several generelizations. Since we intend to study asymptotic phenomena, we assume the symbols to be meromorphic in z; those are holomorphic in the case of differential operators (2.15), while in the pseudo-differential context in expressions for parametrices we have meromorphy, which belongs to the requirements in the cone algebra, to be defined below. The asymptotic types appearing in parametrices, are more general, than the ones above, namely enriched by some finite-dimensional data, suggested by the elliptic theory.

To this end we first define discrete asymptotic types P were first the finite-dimensional data are ignored. Such a P, associated with weight data $\boldsymbol{g} := (\beta, \Theta)$ for a weight β and a certain half-open weight interval $\Theta := (\theta, 0]$, $-\infty \le \theta < 0$, is defined by a sequence

$$P := \{(p_j, m_j)\}_{j=0,\dots,N}, \tag{2.16}$$

where $N \le \infty$ when $\Theta = -\infty$, and $\mathrm{Re}(p_j)$ tends to $-\infty$ when $N = \infty$. In general we suppose that

$$\pi_{\mathbb{C}} P := \{p_j\}_{j=0,\dots,N} \subset \mathbb{C}$$

is contained in the strip $\{z \in \mathbb{C} : \frac{1}{2} - \beta + \theta < \mathrm{Re}(z) < \frac{1}{2} - \beta\}$, and the integers $m_j + 1$ have the meaning of multiplicities of poles p_j of some meromorphic function.

Definition 2.6 A $u \in \mathcal{K}^{s,\beta}(\mathbb{R}_+)$ is said to have discrete asymptotics at $r = 0$ of type P with singular functions $\omega(r) r^{-p_j} \log^k(r)$, if for every $\delta \in \mathbb{R}$ with $\delta > \beta$ we have

$$u(r) - \omega \sum_j \sum_{k=0}^{m_j} c_{jk} r^{-p_j} \log^k(r) \subseteq \mathcal{K}^{s,\delta}(\mathbb{R}_+),$$

for some coefficients $c_{jk} \in \mathbb{C}$, with $\sum_j$ indicating summation over those j, such that $\frac{1}{2} - \delta < \mathrm{Re}(p_j) < \frac{1}{2} - \beta$. We then write $u \in \mathcal{K}_P^{s,\beta}(\mathbb{R}_+)$.

We now analyze the singular functions $\omega(r) r^{-p} \log^k(r)$ in more detail. First note, that

$$\psi(z) := (M\omega)(z) = \frac{1}{z} M(-r \frac{\partial}{\partial r} \omega)(z)$$

is a meromorphic function with a simple pole at $z = 0$. Because of $v(z) := -r \frac{\partial}{\partial r} \omega(r) \in C_0^\infty(\mathbb{R}_+)$ we have $(Mv)(z) \in \mathcal{A}(\mathbb{C})$ and $(Mv)(z)|_{\Gamma_\delta} \in \mathcal{S}(\Gamma_\delta)$ for all δ, uniformly in compact δ-intervals. Furthermore, we have

$$M(r^{-p} \omega(r))(z) = (M\omega)(z - p)$$

with M being interpreted as weighted Mellin transform with sufficiently large p-dependent weight. Differentiating the latter relation with respect to p gives us that the Mellin transform of $\omega(r) r^{-p} \log^k(r)$ is a meromorphic function with a single pole at p of multiplicity $k + 1$. A simple calculation (see [36], Section 2.3.5) yields

$$M(\omega(r) r^{-p} \log^k(r)) = \frac{(-1)^k k!}{(z - p)^{(k+1)}} + a_k(z - p) \tag{2.17}$$

for some function $a_k \in \mathcal{A}(\mathbb{C})$.

Let us now consider asymptotic data in connection of functions in $\mathcal{K}^{s,\beta}(X^{\wedge})$ for $X \in \mathfrak{M}_0$ of dimension n.

To this end we first define discrete asymptotic types P. Such a P, associated with weight data $\boldsymbol{g} := (\beta, \Theta)$ for a weight β and a certain half-open weight interval $\Theta := (\theta, 0]$, $-\infty \leq \theta < 0$, is defined by a sequence of triples

$$P := \{(p_j, m_j, M_j)\}_{j=0,\ldots,N}, \tag{2.18}$$

where $N \leq \infty$ when $\Theta = -\infty$, and $\mathrm{Re}(p_j)$ tends to $-\infty$ when $N = \infty$. In general we suppose that

$$\pi_{\mathbb{C}} P := \{p_j\}_{j=0,\ldots,N} \subset \mathbb{C}$$

is contained in the strip $\{z \in \mathbb{C} : \frac{n+1}{2} - \beta + \theta < \mathrm{Re}(z) < \frac{n+1}{2} - \beta\}$. The meaning of p_j, m_j is as before, while the additional data M_j occurring in (2.18) represent finite-dimensional subspaces of $\mathcal{K}^{\infty,\beta}(X^{\wedge})$. The role of the latter definition is only to extend the above notion to higher-dimensional cones with link X. However, the pseudo-differential calculus, which treats such asymptotics of solutions to elliptic equations, requires asymptotic types of a similar structure also for the involved Mellin symbols, occurring in parametrices of elliptic differential operators of Fuchs type. Those are meromorphic, in contrast to Mellin symbols for differential operators belonging to $M_O^{\mu}(X)$ for a certain μ. The pattern of poles together with multiplicities and coefficient-data represent asymptotic types belonging to meromorphic Mellin symbols that we denote by

$$Q = \{(q_j, n_j.L_j)\}_{j=0,\ldots,N}.$$

This time, the poles $q_j \in \mathbb{C}$ of multiplicity $n_j + 1$ are situated on both sides of some reference weight line $\Gamma_{\frac{n+1}{2}-\beta}$, forming a set $\pi_{\mathbb{C}}(Q)$. Otherwise the sequence of pairs $\{(q_j, n_j)\}_{j=0,\ldots,N}$ is assumed to satisfy similar requirements as those concerning (2.16). Now $M_Q^{\mu}(X)$ is defined to be the space of meromorphic Mellin symbols $l(z)$ with values in $L_{\mathrm{cl}}^{\mu}(X)$ of order μ, with Laurent-coefficients at q_j being finite-dimensional subspaces of $C^{\infty}(X \times X)$.

Observe, that the kernel cut-off operator $H(\psi)$ induces a continuous map

$$H(\psi) : M_Q^{\mu}(X) \to M_O^{\mu}(X)$$

for every μ, and we have $M_Q^{\mu}(X) = M_O^{\mu}(X) + M_Q^{-\infty}(X)$.

Definition 2.7 Let $\mathbb{M}$ be the stretched version of some compact manifold M with conical singularity p. The cone algebra $\{L_{cone}^{\mu}(M; \boldsymbol{g})\}_{\mu,\beta\in\mathbb{R}}$ on $\mathbb{M}$ for weight data $\boldsymbol{g} = (\beta, \beta - \mu, \Theta)$, and real μ, β is defined to be the set of all operators

$$A = r^{-\mu}\omega(r)\mathrm{op}_M^{\beta}(f) + (1 - \omega(r))S + g, \tag{2.19}$$

for some cut-off function $\omega(r)$ supported in a neighbourhood of $r = 0$, a Mellin symbol $f(r, z) \in C^{\infty}(\overline{\mathbb{R}}_+, M_Q^{\mu}(X)$ for some Mellin asymptotic type Q, an element $S \in L_{\mathrm{cl}}^{\mu}(M \setminus \{p\})$, and a so-called Green operator $g \in L_{Green}(M, g)$ to be defined below.

Let M be manifold with conical singularity $\{p\}$, and consider $L^{\mu}(M; g)$, the space of all elements of the form (2.19). Every such A induces continuous operators

$$A : \mathcal{H}^{s,\beta}(M) \to \mathcal{H}^{s-\mu,\beta-\mu}(M), \tag{2.20}$$

for every $s \in \mathbb{R}$. By definition an operator $g \in L_{Green}(M, g)$ is asked to induce continuous operators

$$g : \mathcal{H}^{s,\beta}(M) \to \mathcal{H}_P^{\infty,\beta-\mu}(M), g^* : \mathcal{H}^{s,\beta-\mu}(M) \to \mathcal{H}_Q^{\infty,\beta}(M)$$

for certain asymptotic types P, Q, for every $s \in \mathbb{R}$, with $*$ indicating the adjoint of the respective operator.

Remark 2.8 By definition we have $L^{\mu}(M; g) \subset L_{\mathrm{cl}}^{\mu}(M \setminus \{p\})$. Thus we have the standard principal symbolic structure $\sigma^{\mu}(A)$ of A as an element of $L_{\mathrm{cl}}^{\mu}(M \setminus \{p\})$. In addition the singular stratum $\{p\}$ gives rise to a so-called leading conormal symbol of A namely $\sigma_{cone}^{\mu}(A)(z) := r^{-\mu}f(0, z)$, referring to the conical singularity $\{p\}$, which is operator-valued, namely,

$$\sigma_{cone}^{\mu}(A)(z) := f(0, z) : H^s(X) \to H^{s-\mu}(X).$$

The following result will be mentioned for completeness.

Theorem 2.9 *Let*

$$A \in L^{\mu}(M; g), B \in L^{\nu}(M; h)$$

for $g = (\beta - \nu, \beta - (\mu + \nu), \Theta)$, $h = (\beta, \beta - \nu, \Theta)$ *implies* $AB \in L^{\mu+\nu}(M; c)$, *for* $c = (\beta, \beta - (\mu + \nu), \Theta)$ *and we have* $\sigma^{\mu+\nu}(AB) = \sigma^{\mu}(A)\sigma^{\nu}(B)$, *and* $\sigma_{cone}^{\mu+\nu}(AB)(z) = \sigma_{cone}^{\mu}(A)(z)\sigma_{cone}^{\nu}(B)(z)$.

Definition 2.10 An element $A \in L^{\mu,\beta}(M, g)$ is said to be elliptic, if it is elliptic on $M \setminus \{p\}$ in the standard sense, e.g., with respect to $\sigma^{\mu}(A)$, and if $\sigma_{cone}^{\mu}(A)(z) := f(0, z) : H^s(X) \to H^{s-\mu}(X)$ is an isomorphism for any $s \in \mathbb{R}$ and for all $z \in \Gamma_{\frac{n+1}{2}-\beta}$.

Then the following result is well-known.

Theorem 2.11 *Let $A \in L^{\mu}(M; \boldsymbol{g})$ be elliptic. Then A induces a Fredholm operator* (2.20) *for every $s \in \mathbb{R}$, and A has a parametrix $A^{-1} \in L^{-\mu}(M; \boldsymbol{g}^{-1})$, e.g., we have $A A^{-1} - \mathrm{id}_{\mathcal{H}^{s-\mu,\beta-\mu}(M)} \in L_{Green}(M, \boldsymbol{g}_R)$ for $g_R = (\beta - \mu, \beta - \mu, (-\infty, 0])$ and $A^{-1} A - \mathrm{id}_{\mathcal{H}^{s,\beta}(M)} \in L_{Green}(M, \boldsymbol{g}_L)$ for $g_L = (\beta, \beta, (-\infty, 0])$.*

Manifolds with Edges

3.1 Manifolds with Edge and Weighted Sobolev Spaces

Our next objectives concern aspects on analysis on manifolds with edges. Here we focus again on elliptic PDOs on such configurations. First we fix some notation on the underlying spaces. By a manifold with edge we understand a pseudo-manifold M with an edge $Y \subset M$ for $Y \in \mathfrak{M}_0$ of dimension q. Usually we assume, that Y is compact, but an exception is the half-space $\overline{\mathbb{R}}_+^{q+1}$ with boundary $Y = \mathbb{R}^q$. The picture is similar to the above-mentioned case of a manifold with conical singularities. Those form an "edge" of dimension 0. We have again a disjoint decomposition of M into the smooth manifold $M_{\mathrm{reg}} := M \setminus M_{\mathrm{sing}}$ and $M_{\mathrm{sing}} := Y$, which is the singular part of M.

Definition 3.1 A manifold M with edge is a topological space containing a subspace $Y \in \mathfrak{M}_0$, such that $M \setminus Y$ is a C^∞-manifold, and there is some compact element $X \in \mathfrak{M}_0$, such that every $y \in Y$ has a neighborhood V in M of the form $\Omega \times X^\Delta$ for some open $\Omega \subset Y$.

Example 5 The Cartesian product

$$X^\Delta \times Y$$

for some link $X \in \mathfrak{M}_0$ is an example of a manifold with edge Y.

A more complex example appears, when we first start from an X-bundle M_1 over a Y and then form an X^Δ-bundle M by fibrewise replacing X by X^Δ.

Example 6 Another example is generated by first looking at $M_{[0,t]}$ which defined as the union of the $\left(X^\Delta \cap ((0, 1] \times X)\right)$-bundle over Y, with the zero-section, identified with Y by $r = 0$, while fibrewise $X^\Delta|_{r \geq 1}$ is replaced by a compact space B, which is a smooth

D.-C. Chang, B.-W. Schulze, *Analysis on Manifolds with Singularities*,
Pseudo-Differential Operators 16, https://doi.org/10.1007/978-3-032-01890-8_3

manifold with boundary X (identified with $\{1\} \times X$). This gives us altogether a manifold M with edge Y, which is different from that of Example 3.1. M is compact, when both Y and B are compact. On a manifold M with edge Y the half-axis $\overline{\mathbb{R}}_+ \ni r$ plays the role of the normal direction of a collar neighborhood of the edge.

Let us now briefly recall some basics on weighted Sobolev spaces on manifolds with edge. This will belong to the background of the corresponding operator calculus, in particular, the so-called edge algebra, see the material below. More details may be found in [68]. Let us first recall a definition of abstract edge spaces, denoted by $\mathcal{W}^s(\mathbb{R}^q, H)$, $s \in \mathbb{R}$, for some separable Hilbert space H.

The definition is based on the action of some one-parameter group $\{\kappa_\iota\}_{\iota \in \mathbb{R}_+}$ of strongly continuous operators on H, satisfying the above-mentioned relations.

We then define

$$\|u\|_{\mathcal{W}^s(\mathbb{R}^q, H)} = \left\{ \int [\eta]^{2s} \|\kappa^{-1}(\eta)(F_{y \to \eta} u)(\eta)\|_H^2 d\eta \right\}^{1/2} \tag{3.1}$$

for $\kappa(\eta) := \kappa_{[\eta]}$, first on $\mathcal{S}(\mathbb{R}^q, H)$ and then by completion with respect to the indicated norm. In particular, we can check, that

$$\mathcal{W}^s(\mathbb{R}^q, H^s(\mathbb{R}^m)) = H^s(\mathbb{R}^{q+m}). \tag{3.2}$$

On a manifold with edge like in Example 3.1 we are interested in spaces $\mathcal{K}^{s,\beta}(X^\triangle)$, and we can form the wedge spaces

$$\mathcal{W}^{s,\beta}(\mathbb{R}^q \times X^\triangle) := \mathcal{W}^s(\mathbb{R}^q, \mathcal{K}^{s,\beta}(X^\triangle)),$$

or in stretched version

$$\mathcal{W}^{s,\beta}(\mathbb{R}^q \times X^\wedge) := \mathcal{W}^s(\mathbb{R}^q, \mathcal{K}^{s,\beta}(X^\wedge)). \tag{3.3}$$

Since we mainly look at the involved spaces $\mathcal{K}^{s,\beta}(X^\wedge)$, referring to $X^\wedge$ (in stretched version), we recall once again the definition of the corresponding spaces. We usually set

$$\mathcal{K}^{s,\beta}(X^\wedge) = \{\omega u + (1 - \omega)v : u \in \mathcal{H}^{s,\beta}(X^\wedge), v \in H^s_{\mathrm{cone}}(X^\wedge)\}, \tag{3.4}$$

where ω is any cut-off function, and $H^s_{\mathrm{cone}}(X^\wedge) := H^s_{\mathrm{cone}}(\mathbb{R} \times X)|_{\mathbb{R}_+ \times X}$, defined to be the set of all $f \in H^s_{\mathrm{loc}}(\mathbb{R} \times X)|_{\mathbb{R}_+ \times X}$, such that for any coordinate neighborhood U of X and any diffeomorphism $\chi_1 : U \to V$ to an open set $V \subset \mathbb{S}^n$ (with $\mathbb{S}^n$ being the unit sphere in $\mathbb{R}^{n+1}$) we have

$$(1 - \omega)\varphi(\chi^{-1})^* f \in H^s(\mathbb{R}^{1+n})$$

for every $u \in C_0^\infty(U)$ and any cut-off function ω. Here $\chi(r,x) := (r, r\chi_1(x))$. The definition is independent of the choice of ω. Let us point out once again, that $H_{\mathrm{cone}}^s(\mathbb{R} \times X)|_{\mathbb{R}_+ \times X}$ indicates an H^s-space, referring to a conical opening of the underlying space $\mathbb{R}_+ \times X \ni (r,x)$ as $r \to \infty$. Clearly, this is the case also for $\mathcal{K}^{s,\beta}(X^\wedge)$. Since this interpretation comes from the behavior of $H^s(\mathbb{R}^{1+n})$ globally in $\mathbb{R}^{1+n}$ for $\tilde{x} \in \mathbb{R}^{1+n}$ and $|\tilde{x}|$ tending to ∞, we just obtain, that $\mathcal{W}^{s,\beta}(\mathbb{R}^q \times X^\wedge) \subset H_{\mathrm{loc}}^s(\mathbb{R}^q \times X^\wedge)$ for every real s, β.

Let us also mention a relation of the conical opening of $X^\wedge$ for $r \to \infty$ in connection with results of H.O. Cordes [18] in connection with manifolds M, which are "asymptotically flat" at ∞. A simplest manifold M of that kind is $\mathbb{R}^{1+n}$. Pseudo-differential operators globally in $\mathbb{R}^{1+n}$ have been studied also by C. Parenti [51], M.A. Shubin [78], and many other authors. Here we refer to material of E. Schrohe [63], in particular, to the spaces $H^{(s_1,s_2)}(\mathbb{R}^{1+n})$ for $s_1, s_2 \in \mathbb{R}$, which are analogues of the usual Sobolev spaces $H^s(\mathbb{R}^{1+n}) = H^{(s,0)}(\mathbb{R}^{1+n})$. $s \in \mathbb{R}$. The space $H^{(s_1,s_2)}(\mathbb{R}^{1+n})$ is defined to be the completion of $\mathcal{S}(\mathbb{R}^{1+n})$ with respect to the corresponding global norm.

3.2   Edge Spaces with Asymptotics

We now pass to the main topic of this section, namely, to subspaces of (3.3) with asymptotics, where ellipticity of elements in the edge algebra should cause regularity of solutions in edge spaces with asymptotics. The method to formulating edge asymptotics is to insert spaces of the kind $\mathcal{K}_P^{s,\beta}(X^\wedge)$ in (3.3) instead of the spaces $\mathcal{K}^{s,\beta}(X^\wedge)$ in (3.3) for certain asymptotic types $P = P(y)$.

This gives spaces of the kind

$$\mathcal{W}_P^{s,\beta}(\mathbb{R}^q \times X^\wedge) := \mathcal{W}^s(\mathbb{R}^q, \mathcal{K}_P^{s,\beta}(X^\wedge))$$

However, the problem with discrete P in this connection is, that those may be variable and branching, depending on the reference point $y \in Y$. Moreover, we have to observe additional edge conditions, analogous to boundary conditions in elliptic boundary problems, when we interpret elliptic boundary problems as particular elliptic edge problems. Nevertheless, special elliptic edge problems will suggest an adequate answer in the general case.

3.3   Edge Operators

Local edge-degenerate operators will appear (up to so-called Green operators with asymptotics and a Mellin operator convention, to be explained in connection with the edge algebra below) in the form

$$A = r^{-\mu} \mathrm{op}_y \mathrm{op}_r(\tilde{p}(r, y, r\rho, r\eta)) \tag{3.5}$$

for operator-valued symbols

$$\tilde{p}(r, y, \tilde{\rho}, \tilde{\eta}) \in C^\infty(\overline{\mathbb{R}}_+ \times \mathbb{R}^q, L^\mu_{\mathrm{cl}}(X; \mathbb{R}^{1+q}_{\tilde{\rho}, \tilde{\eta}})). \tag{3.6}$$

The subclass of local edge-degenerate differential operators is defined by replacing the latter symbol class by

$$\tilde{p}(r, y, \tilde{\rho}, \tilde{\eta}) \in C^\infty(\overline{\mathbb{R}}_+ \times \mathbb{R}^q, \mathrm{Diff}^\mu(X; \mathbb{R}^{1+q}_{\tilde{\rho}, \tilde{\eta}})). \tag{3.7}$$

with $\mathrm{Diff}^\mu(X; \mathbb{R}^{1+q}_{\tilde{\rho}, \tilde{\eta}})$ being the space of parameter-depending differential operators of order $\mu \in \mathbb{N}$ with parameters $(\tilde{\rho}, \tilde{\eta}) \in \mathbb{R}^{1+q}_{\tilde{\rho}, \tilde{\eta}}$, with subsequent substitution $(\tilde{\rho}, \tilde{\eta}) = (r\rho, r\eta)$. Note that the Laplace-Beltrami operator belonging to the wedge metric of $\mathbb{R}^q \times X^\Delta$ is of this kind, for $\mu = 2$.

Example 7 Consider an edge-degenerate differential operator

$$A = r^{-\mu} \sum_{j + |\alpha| \leq \mu} a_{j\alpha}(r, y)\left(-r\frac{\partial}{\partial r}\right)^j (rD_y)^\alpha \tag{3.8}$$

with $a_{j\alpha}(r, y) \in C^\infty(\overline{\mathbb{R}}_+ \times \Omega, \mathrm{Diff}^{\mu-(j+|\alpha|)}(X))$, $\Omega \subseteq \mathbb{R}^q$ open. The operator (3.8) can also be written in the form

$$A = r^{-\mu} \mathrm{op}_M \mathrm{op}_y(h) \tag{3.9}$$

for

$$h(r, y, z, r\eta) = \sum_{j + |\alpha| \leq \mu} a_{j\alpha}(r, y)z^j (r\eta)^\alpha$$

Locally in $(r, x, y) \in \mathbb{R}_+ \times \Sigma \times \Omega$ for some coordinate neighborhood $\Sigma \subset X$ the complete symbol of (3.8) takes the form

$$p_\psi(r, x, y, \rho, \xi, \eta) = r^{-\mu} \sum_{j + |\alpha| \leq \mu} \sum_{|\beta| \leq \mu - (j + |\alpha|)} b_{j\alpha\beta}(r, x, y)\xi^\beta(-ir\rho)^j (r\eta)^\alpha \tag{3.10}$$

with coefficients $b_{j\alpha\beta}(r, x, y) \in C^\infty(\overline{\mathbb{R}}_+ \times \Sigma \times \Omega)$. We also write

$$p(r, y, \rho, \eta) := r^{-\mu} \sum_{j + |\alpha| \leq \mu} \sum_{|\beta| \leq \mu - (j + |\alpha|)} b_{j\alpha\beta}(r, x, y)D_x^\beta(-ir\rho)^j (r\eta)^\alpha.$$

Note, that edge-degenerate operators A have two principal symbolic levels. Those will determine ellipticity and the operator calculus including parametrices.

In other words, we have

$$\left(\sigma_\psi^\mu(A)(t, x, y, \rho, \xi, \eta), \sigma_\wedge^\mu(A)(y, \eta)\right), \tag{3.11}$$

with $\sigma_\psi^\mu(A)$ being the homogeneous principal part of (3.10), while the principal edge symbol is given by

$$\sigma_\wedge^\mu(A)(y, \eta) = r^{-\mu}\mathrm{op}_r(\tilde{p}(0, y, r\rho, r\eta)).$$

It is a (y, η)-dependent operator family

$$\sigma_\wedge^\mu(A)(y, \eta) : \mathcal{K}^{s,\beta}(X^\wedge) \to \mathcal{K}^{s-\mu,\beta-\mu}(X^\wedge) \tag{3.12}$$

from the parameter-dependent cone calculus on the (stretched) cone, transversally to the (local) edge $\mathbb{R}^q$, see more explanations below. Relation (3.12) holds for differential operators. However, in the pseudo-differential case we will employ the following version of Mellin pseudo-differential calculus, adapted to formulations of the edge algebra. (also called "Mellin quantization").

Theorem 3.2 *Consider a symbol* $p(r, y, \rho, \eta) \;:=\; \tilde{p}(r, y, \tilde{\rho}, \tilde{\eta})|_{\tilde{\rho}=r\rho, \tilde{\eta}=r\eta}$ *for* $\tilde{p}(r, y, \tilde{\rho}, \tilde{\eta}) \in C^\infty(\overline{\mathbb{R}}_+ \times \Omega, L_{\mathrm{cl}}^\mu(X; \mathbb{R}_{\tilde{\rho},\tilde{\eta}}^{1+q}))$, *where* $\Omega \subseteq \mathbb{R}^q$ *is an open set. Then there is an* $h \in C_{\mathrm{deg}}^\infty(\Omega, M_O^\mu(X; \mathbb{R}^q))$, *where* $h(r, z, \eta) = \tilde{h}(r, z, r\eta)$, *such that*

$$\mathrm{op}_r(p)(y, \eta) - \mathrm{op}_M^\beta(h)(y, \eta) = \mathrm{op}_r(Q)(y, \eta) \tag{3.13}$$

with $Q(r, r', y, \rho, \eta) = (1 - \psi(r'/r))p(r, y, \rho, \eta)$, *for some* $\psi \in C_0^\infty(\mathbb{R}_+)$ *such that* $\psi \equiv 1$ *near* 1.

The correspondence $p \to h$ is sometimes denoted by Q, see also [41] or [20]. Note that Q also has an inverse, due to the above-mentioned equivalence of the respective phase functions.

In connection with Theorem 3.2 we now introduce so-called edge symbols of order μ, namely,

$$a(y, \eta) = \sigma_1(r)\{a_0(y, \eta) + a_1(y, \eta)\}\sigma_0(r) + g(y, \eta), \tag{3.14}$$

where

$$a_0(y, \eta) = \omega_1(r[\eta])r^{-\mu}\mathrm{op}_M^{\beta-n/2}h(y, \eta)\omega_0(r[\eta]),$$

$$a_1(y, \eta) = (1 - \omega_1(r[\eta])r^{-\mu}\mathrm{op}_r(p)(y, \eta)(1 - \omega_2(r[\eta]),$$

with Green symbols $g(y, \eta)$. Expressions of the form (3.14) play the role of upper left corners of edge symbols

$$a(y, \eta) : \mathcal{K}^{s,\beta}(X^\wedge) \to \mathcal{K}^{s-\mu,\beta-\mu}(X^\wedge), \tag{3.15}$$

when p and h are associated with each other via relation (3.13). Green remainders $g(y, \eta)$ are families of compact operators in the edge calculus. According to results of [28] expression (3.14) admits an equivalent representation

$$\sigma_\wedge^\mu(A)(y, \eta) = r^{-\mu}\mathrm{op}_M^{\beta-\frac{n}{2}}(h_\wedge)(y, \eta) + g(y, \eta) \tag{3.16}$$

for another Green symbol $g(y, \eta)$ and $h_\wedge(r, z, \eta) := \tilde{h}(0, z, r\eta)$ for $A := \mathrm{op}_y(a)$. The nature of Green symbols in the edge calculus is also characterizised in [28], so we may refer to it in the present exposition. Let us now give the definition of upper left corners of the edge algebra.

Definition 3.3 Let $\mathbb{W}$ be the stretched version of some compact (pseudo)-manifold W with edge Y. The space $\{L_{edge}^\mu(W; \boldsymbol{g})\}_{\mu,\beta\in\mathbb{R}}$ of upper left corners of the edge algebra on $\mathbb{W}$ for some weight data $\boldsymbol{g} = (\beta, \beta - \mu, \Theta)$ and any real μ, β is defined to be the set of all operators

$$A = \omega(r) \sum_{j=1}^N \varphi_j(y)\left[\mathrm{op}_y r^{-\mu}\omega(r[\eta])\mathrm{op}_M^{\beta-\frac{n}{2}}(h_{j,\wedge})(y, \eta) + g_j(y, \eta)\right] + (1 - \omega(r))S,$$
$$\tag{3.17}$$

where $\{\varphi_j\}_{j=1,...,N}$ is a partition of unity, subordinated to an open covering $\{U_j\}_{j=1,...,N}$ of Y, with $h_{j,\wedge}$ and g_j being related to U_j in the sense of (3.16), and some $S \in L_{\mathrm{cl}}^\mu(W \setminus Y)$.

We have $L_{edge}^\mu(W; \boldsymbol{g}) \subset L_{\mathrm{cl}}^\mu(W \setminus Y)$, such for any A in that space we have a homogeneous principal symbol of order μ

$$\sigma_\psi^\mu(A)(r, x, y, \rho, \xi, \eta). \tag{3.18}$$

In addition, (3.16) induces continuous operators

$$\sigma_\wedge^\mu(A)(y, \eta) : \mathcal{K}^{s,\beta}(X^\wedge) \to \mathcal{K}^{s-\mu,\beta-\mu}(X^\wedge), \tag{3.19}$$

which is a family of cone operators on the infinite stretched cone $X^\wedge$.

Edge Operators with Trace and Potential Conditions $\quad$ 4

4.1 $\quad$ Basic Tools

This consideration is devoted to an elementary approach to pseudo-differential operators on singular manifolds. We partly come back to considerations of previous parts of this exposition, under modified notation. After some years of development of this area and remarkable contributions by numerous authors, stimulated by classical and new applications, the field became somehow heterogeneous and complicated. It is therefore desirable, to point out elementary notions and methods, that admit a reorganisation of structures and a more convenient access.

Starting point are some observations on Sobolev spaces $H^s(\mathbb{R}^m)$, for $s \in \mathbb{R}$ and $m \in \mathbb{N}$ ($\mathbb{N} := \{0, 1, 2, \ldots\}$). We have $H^0(\mathbb{R}^m) = L^2(\mathbb{R}^m)$ with the standard space of complex square integrable functions in $\mathbb{R}^m$, with the scalar product

$$\langle u, v \rangle_{L^2(\mathbb{R}^m)} = \int_{x \in \mathbb{R}^m} u(x)\overline{v}(x)dx \, . \tag{4.1}$$

We refer to the identity (Parseval's formula)

$$\langle u, v \rangle_{L^2(\mathbb{R}^m)} = (2\pi)^{-m} \int_{\xi \in \mathbb{R}^m} (Fu)(\xi)(\overline{Fv})(\xi)d\xi \, , \tag{4.2}$$

where

$$F : L^2(\mathbb{R}^m) \to L^2(\mathbb{R}^m), u \mapsto \left(\xi \mapsto \int u(x)e^{-i\langle x, \xi \rangle}dx \right)$$

© Springer Nature Switzerland AG 2025
D.-C. Chang, B.-W. Schulze, *Analysis on Manifolds with Singularities*,
Pseudo-Differential Operators 16, https://doi.org/10.1007/978-3-032-01890-8_4

is the Fourier transform. Sobolev spaces are defined (up to equivalence of norms) to be the completion of $\mathcal{S}(\mathbb{R}^m)$ (the Schwartz space in $\mathbb{R}^m$) with respect to the norm

$$\|u\|_{H^s(\mathbb{R}^m)} = \left(\int_{\xi \in \mathbb{R}^m} (\langle\xi\rangle^{2s} |(Fu)(\xi)|^2 d\xi \right)^{1/2}, \tag{4.3}$$

for $\langle\xi\rangle := (1 + |\xi|^2)^{1/2}$. Sometimes it is useful, to pass to equivalent norms ($\approx$, later also denoted by $=$). For Sobolev spaces on $\mathbb{R}^{n+1+q}$, $q \in \mathbb{N}$, we will take, e.g.,

$$\|w\|_{H^s(\mathbb{R}^{n+1+q})} = \left(\int_{(\tilde\xi,\eta)\in\mathbb{R}^{m+q}} \langle(\tilde\xi,\eta)\rangle^{2s} |(Fw)(\tilde\xi,\eta)|^2 d\tilde\xi d\eta \right)^{1/2}$$

$$\approx \left(\int_{(\tilde\xi,\eta)\in\mathbb{R}^{m+q}} (\langle\tilde\xi\rangle^2 + [\eta]^2)^s |(Fw)(\tilde\xi,\eta)|^2 d\tilde\xi d\eta \right)^{1/2}$$

$$\approx \left(\int_{(\tilde\xi,\eta)\in\mathbb{R}^{m+q}} (|\tilde\xi|^2 + [\eta]^2)^s |(Fw)(\tilde\xi,\eta)|^2 d\tilde\xi d\eta \right)^{1/2}, \tag{4.4}$$

where $m := 1 + n$ and $\tilde\xi := (\rho, \xi) \in \mathbb{R}^{1+n}$, with $[\eta]$ being a strictly positive smooth function, which is equal to $|\eta|$, when $|\eta| > 1$.

We will interpret η as a local edge-covariable, belonging to the "edge" $\mathbb{R}^q \ni y$. For the special case $q = 0$ we obtain again (4.3). The Fourier transform F is denoted by the same letter for different dimensions. When we indicate the transformation of variables, we also write $F_{x\to\xi}$ (e.g., when F refers to $\mathbb{R}^n := \mathbb{R}^n_x$).

We now rephrase (4.4) in terms of

$$\kappa(\eta) := \kappa_{[\eta]}, \tag{4.5}$$

which belongs to the motivation of spaces with values in another Hilbert space E, here for the case $E = H^s(\mathbb{R}^m)$ with $(\kappa_\lambda f)(\tilde x) := \lambda^{m/2} f(\lambda\tilde x)$. Note that

$$\kappa_\lambda(F_{\tilde x\to\tilde\xi} f)(\tilde\xi) = F_{\tilde x\to\tilde\xi}(\kappa_\lambda^{-1} f)(\tilde\xi). \tag{4.6}$$

Lemma 4.1 *For every $s \in \mathbb{R}$ we have (up to equivalence of norms)*

$$\|u\|^2_{H^s(\mathbb{R}^{n+1+q})} = \iint (|\tilde\xi|^2 + [\eta]^2)^s |(Fu)(\tilde\xi,\eta)|^2 d\tilde\xi d\eta$$

$$= \iint [\eta]^{2s} (1 + (\frac{|\tilde\xi|}{[\eta]})^2)^s |(Fu)(\tilde\xi,\eta)|^2 d\tilde\xi d\eta$$

$$= \iint [\eta]^{2s} (1 + |\tilde\xi|^2)^s |(Fu)([\eta]\tilde\xi,\eta)|^2 [\eta]^{n+1} d\tilde\xi d\eta$$

$$= \int [\eta]^{2s} \int (1 + |\tilde{\xi}|^2)^s |\kappa(\eta)(Fu)(\tilde{\xi}, \eta)|^2 d\tilde{\xi} d\eta.$$

Note that the above computation can be continued by

$$\|u\|^2_{H^s(\mathbb{R}^{n+1+q})} = \int [\eta]^{2s} \left\| \kappa^{-1}(\eta)(F_{y\to\eta}u)(\tilde{x}, \eta) \right\|^2_{H^s(\mathbb{R}^{n+1}_{\tilde{x}})} d\eta$$

$$= \left\| (F_{y\to\eta}\kappa^{-1}(\eta)u)(\tilde{x}, \eta) \right\|^2_{H^s(\mathbb{R}^q_y, H^s(\mathbb{R}^{n+1}_{\tilde{x}}))}.$$

In other words, Lemma 4.1 consists of an elementary substitution in a norm expression for Sobolev spaces in Euclidean space for any real s. The role of the right-hand side of the latter relation is to interpret $E := H^s(\mathbb{R}^{n+1})$ as an example of a space on $\mathbb{R}^{n+1}$, where the origin $0 \in \mathbb{R}^{n+1}$ is regarded as a conical singularity with link $\mathbb{S}^n$, which is just the unit sphere in $\mathbb{R}^{n+1}$. Here $H^s(\mathbb{R}^{n+1+q})$ itself is formally an example of an edge space, see further details below, e.g., from [68], with an involved group action $(\kappa_\lambda)_{\lambda \in \mathbb{R}_+}$ on the space E.

Remark 4.2 A modification of Lemma 4.1 will be formulated below for edge Sobolev spaces $\mathcal{W}^s(\mathbb{R}^q, E)$ over the edge $\mathbb{R}^q$ instead of $H^s(\mathbb{R}^{n+1+q})$, where $H^s(\mathbb{R}^{n+1})$ is replaced by the space

$$E = \mathcal{K}^{s,\beta}(\mathbb{R}_+ \times X), \tag{4.7}$$

defined in formula (4.21) below for $\beta \in \mathbb{R}$, which is also endowed with a group action, with X being the link of dimension n belonging to a manifold with conical singularity, e.g., $X = \mathbb{S}^n$, the unit spere in $\mathbb{R}^{n+1}$.

We will see, that the intention of this construction is to avoid contradictions in the definition of edge Sobolev spaces globally on a manifold B with edge Y via localizations close to Y and far from Y, obtained by glueing together the respective local terms with the help of a partition of unity on B.

The nature of edge spaces may also be illustrated by the following observations on standard Sobolev spaces. First note, that the operator of restriction

$$r_0 : \mathcal{S}(\mathbb{R}^m) \to \mathbb{C}$$

defined by $r_0 f = f(0)$ induces a continuous extension

$$r_0 : H^s(\mathbb{R}^m) \to \mathbb{C} \tag{4.8}$$

for every $s > m/2$. In fact, writing $u(0) = \int \hat{u}(\xi)d\xi$ for $\hat{u}(\xi) = (F_{x\to\xi}u)(\xi)$, i.e., $u(x) = \int e^{ix\xi}\hat{u}(\xi)d\xi$, we obtain $|u(0)| \le \int |\hat{u}(\xi)|d\xi$ which is $\le$

$$\int \langle\xi\rangle^{-s}\langle\xi\rangle^s|\hat{u}(\xi)|d\xi \le \{\int \langle\xi\rangle^{2s}|\hat{u}(\xi)|^2 d\xi \int \langle\xi\rangle^{-2s}d\xi\}^{1/2},$$

and the latter expression is $\le c\|u\|_{H^s(\mathbb{R}^m)}$ for $c^2 = \int \langle\xi\rangle^{-2s}d\xi$. This yields altogether $|u(0)| \le c\|u\|_{H^s(\mathbb{R}^m)}$ for $s > m/2$. The map (4.8) may be regarded as an operator-valued symbol $r_0 : E \to \tilde{E}$ for the spaces $E = H^s(\mathbb{R}^m)$ and $\tilde{E} = \mathbb{C}$, satisfying the relation

$$(\kappa_\lambda u)(x) = \lambda^{\frac{m}{2}}u(\lambda x) \tag{4.9}$$

for all $\lambda \in \mathbb{R}_+, u \in H^s(\mathbb{R}^m)$. This entails

$$\mathcal{W}^s(\mathbb{R}^q, H^s(\mathbb{R}^m)) = H^s(\mathbb{R}^{q+m}). \tag{4.10}$$

In fact, we have as before $\|u\|_{H^s(\mathbb{R}^m)} = \{\int \langle\xi\rangle^{2s}|(F_{x\to\xi}u)(\xi)|^2 d\xi\}^{1/2}$, and $\|f\|_{H^s(\mathbb{R}^{m+q})} = \{\int\int(|\xi|^2 + [\eta]^2)(Ff)(\xi, \eta)|^2 d\xi d\eta\}^{1/2}$, which gives us relation (4.10). Recall, that $\mathbb{R}^q$ may be interpreted as a q-dimensional edge of $\mathbb{R}^{q+m}$. This suggests elliptic theory of pseudo-differential operators in Sobolev spaces over $\mathbb{R}^{q+m}$ with trace- and potential conditions referring to Sobolev data over $\mathbb{R}^q$, expressed by operator-valued symbols, analogously to elliptic theory of boundary value problems on the half-space $\mathbb{R}_+^{q+1}$ with respect to the boundary $\mathbb{R}^q$, known from Boutet de Monvel's theory (localized near the boundary). Let us sketch some detail connected with this idea, which entails at the same time a remark on anisotropic descriptions of standard Sobolev spaces as well on certain transmission problems, to be specified below. Assume, we are given a classical pseudo-differential operator A in $\mathbb{R}^{q+m}$ of order $\mu \in \mathbb{R}$, which induces continuous operators $A : H^s(\mathbb{R}^{q+m}) \to H^{s-\mu}(\mathbb{R}^{q+m})$ for all $s \in \mathbb{R}$. For such a property globally in $\mathbb{R}^{q+m}$ we assume for simplicity, that the complete symbol $a(y, x, \eta, \xi)$ of A is independent of (y, x) for large $|(y, x)|$ and also, that a possible smoothing remainder vanishes. Let $a^{(\mu)}(y, x, \eta, \xi)$ denote the homogeneous principal symbol of A. Assume ellipticity, i.e., $a^{(\mu)}(y, x, \eta, \xi) \ne 0$ for $(\eta, \xi) \ne 0$. We now consider the family of operators

$$\sigma_\partial^{(\mu)}(A)(y, \eta) := \mathrm{op}_x\left(a^{(\mu)}(y, 0, \eta, \xi)\right) : H^s(\mathbb{R}^m) \to H^{s-\mu}(\mathbb{R}^m) \tag{4.11}$$

for $\eta \ne 0$.

Lemma 4.3 *We have*

$$\sigma_\partial^{(\mu)}(A)(y, \lambda\eta) = \lambda^\mu \kappa_\lambda \sigma_\partial^{(\mu)}(A)(y, \eta)\kappa_\lambda^{-1} \tag{4.12}$$

for $\lambda > 0$.

Proof Relation (4.12) may be verified as follows. We have

$$\sigma_\partial^{(\mu)}(A)(y, \lambda\eta)u(x) = \iint e^{i(x-x')\cdot\xi} a^{(\mu)}(y, 0, \lambda\eta, \xi)u(x')dx'đ\xi$$

for $u \in H^s(\mathbb{R}^m)$, or

$$\sigma_\partial^{(\mu)}(A)(y, \lambda\eta)u(x) = F_{\xi\to x}^{-1} a^{(\mu)}(y, 0, \lambda\eta, \xi) F_{x'\to\xi} u(x').$$

Using the homogeneity

$$a^{(\mu)}(\ldots, \lambda\eta, \lambda\lambda^{-1}\xi) = a^{(\mu)}(\ldots, \lambda\eta, \lambda\xi') = \lambda^\mu a^{(\mu)}(\ldots, \eta, \xi')$$

it follows, that

$$\sigma_\partial^{(\mu)}(A)(y, \lambda\eta)u(x) = \lambda^\mu F_{\xi\to x}^{-1} \lambda^\mu a^{(\mu)}(y, 0, \eta, \lambda^{-1}\xi) \int e^{-ix'\cdot\xi} u(x')dx'.$$

Substituting in the latter formula $\xi' = \lambda^{-1}\xi$ we obtain

$$\sigma_\partial^{(\mu)}(A)(y, \lambda\eta)u(x) = \lambda^\mu F_{\xi\to x}^{-1} a^{(\mu)}(y, 0, \eta, \xi') \int e^{-i\lambda x'\cdot\xi'} u(x')dx',$$

which gives us for $x'' = \lambda x'$ (up to powers of λ which may be dropped)

$$\sigma_\partial^{(\mu)}(A)(y, \lambda\eta)u(x) = \lambda^\mu \int e^{i\xi'\cdot\lambda x} a^{(\mu)}(y, 0, \eta, \xi') \int e^{-ix''\cdot\xi'} u(\lambda^{-1}x'')dx''đ\xi'.$$

Thus, according to (4.12) we get

$$\sigma_\partial^{(\mu)}(A)(y, \lambda\eta)u(x) = \lambda^\mu \kappa_\lambda F_{\xi\to x}^{-1} a^{(\mu)}(y, 0, \eta, \xi') F_{x''\to\xi'} \kappa_\lambda^{-1} u(x').$$

Due to formula (4.9) we could replace $\kappa_\lambda u(x)$ by $u(\lambda x)$. $\square$

In other words, we obtained

$$\operatorname{op}_x(a)(y, \eta) \in S^\mu(\mathbb{R}_y^q \times \mathbb{R}_\eta^q; H^s(\mathbb{R}^m), H^{s-\mu}(\mathbb{R}^m))$$

which just asserts, that $\operatorname{op}_x(a)(y, \eta)$ belongs to the space of operator-valued edge symbols $S^\mu(\mathbb{R}_y^q \times \mathbb{R}_\eta^q; E, \tilde{E})$ for $E := H^s(\mathbb{R}^m)$, $\tilde{E} := H^{s-\mu}(\mathbb{R}^m)$, defined by the edge symbolic estimates, associated with the group action, mentioned in relation (4.9). A similar relation may be proved, when we admit a symbol of the form $a(y, \eta, x, \xi)$ rather than $a(y, \eta, 0, \xi)$, when we assume, for convenience, that the symbol a is independent of x for $|x|$ sufficiently large. The technique for this extension contains a tensor product argument, which allows us to interpret $a(y, \eta, x, \xi)$ as an element of $a(y, \eta, c, \xi)\hat{\otimes}_\pi C_0^\infty(\mathbb{R}_x^n)$ for some constant $c \in \mathbb{R}_x^n$. A slight generalization allows us to generalize the consideration to symbols $a(y, \eta, x, \xi)$, also depending on the variable x for the proof, when the symbol a contains an extra dependence on x. Then the ellipticity of $a(y, \eta, x, \xi)$ shows that

$$\mathrm{op}_x(a)(y, \eta) : H^s(\mathbb{R}^m) \to H^{s-\mu}(\mathbb{R}^m)$$

is a family of Fredholm operators between the involved spaces, for $\eta \neq 0$. The same is true of (4.11).

4.2   Group Actions and Weighted Cone and Edge Spaces

The above-mentioned group action $(\kappa_\lambda)_{\lambda \in \mathbb{R}_+}$ acting on suitable spaces E gives rise to new classes of distribution spaces, occurring in the analysis on manifolds with singularities. Let $\mathfrak{M}_0$ be a category of smooth compact and oriented manifolds, endowed with some Riemannian metric. This will be the starting point of other geometric objects, called singular manifolds, to be studied below. Those will form categories $\mathfrak{M}_k$, where $k \in \mathbb{N}$ has the meaning of an involved singularity order, e.g., for $k = 1$ of conical or edge type (precise definitions and examples will be given below). General $k > 1$ will indicate a corner singularity, e.g., in the case $\overline{\mathbb{R}}_+^k$ with $\overline{\mathbb{R}}_+$ being the closed half-axis regarded as an infinite cone with the origin as conical singularity. Another example of a conical singularity is a point v on a C^0-manifold D of dimension $n + 1$ for $n \in \mathbb{N}$, when $v \in D$ and $D \setminus \{v\}$ is smooth, and there is a neighborhood $V \subset D$ of v with smooth boundary such that V can be identified with the union of straight connections of the origin in $\mathbb{R}^{N+1}$ (identified with v and for sufficiently large $N \geq n$) with all points of ∂V identified with points in $\mathbb{S}^N$, the unit sphere in $\mathbb{R}^{N+1}$. In other words, there is a map $\phi : V \to B^{N+1}$ (with B^{N+1} being the unit ball in $\mathbb{R}^{N+1}$ centered at the origin 0), such that $\phi(\partial V) \to \mathbb{S}^N$ is an embedding into the unit sphere $\mathbb{S}^N$ in $\mathbb{R}^{N+1}$ and $\phi(v) = 0$. Moreover, $\phi(V)$ is the union $K \subset B^{N+1}$ of all straight lines, connecting 0 with all points of $\phi(\partial V)$, such that $\phi : V \to K$ is altogether a bijection. We call $X := \partial V$ the link of the conical singularity v. By construction we have $X \in \mathfrak{M}_0$.

Another definition of a conical singularity may be found in [65], page 101, concerning the case of manifolds with boundary.

It is common in this context, instead of the conical singularity, to speak of the stretched manifold, i.e., similar to r, x-variables with $r \in \mathbb{R}_+$ and $x \in X$, occurring in polar coordinates close to v, to consider a cylindrical end, like $\overline{\mathbb{R}}_+ \times X := \mathbb{V}$ instead of the above V. This has the advantage to avoid a clumsy definition of how the Riemannian metric is degenerate near v. Therefore, we tacitly speak of the corresponding stretched manifold $\mathbb{D}$ in (r, x)-coordinates near the stretched conical singularity, see also [48], or [47]. This is a quite classical point of view, even if the origin $0 \in \mathbb{R}^{N+1}$ is a fictitious conical singularity, since the geometry across 0 is smooth, namely to see partial differential operators in $\mathbb{R}^{N+1}$ with smooth coefficients in polar coordinates. The corresponding calculation may be found, e.g., in [65], and it says, that a differential operator of order $\mu \in \mathbb{N}$ in the variables $\tilde{x} \in \mathbb{R}^{n+1}$

$$\tilde{A} := \sum_{|\tilde{\alpha}| \leq \mu} a_{\tilde{\alpha}}(\tilde{x}) D_{\tilde{x}}^{\tilde{\alpha}} \tag{4.13}$$

takes the form of an operator of Fuchs type, referring to the origin of $\mathbb{R}^{n+1}$, regarded as conical singularity, with the link $X := \mathbb{S}^n$. In fact, introducing polar coordinates in (4.13) gives us

$$A = r^{-\mu} \sum_{j+|\alpha| \leq \mu} a_{j,\alpha}(r) \left(r \frac{\partial}{\partial r} \right)^j, \tag{4.14}$$

for coefficients $a_{j,\alpha}(r) \in C^\infty(\overline{\mathbb{R}}_+, \mathrm{Diff}^{|\alpha|}(X))$, with $\mathrm{Diff}^k(X)$ being the space of partial differential operators of order k on X in its natural Fréchet topology. Conversely, examples of operators of the form (4.14) for the case $X := \mathbb{S}^1$ show, that the space of operators (4.14) is much larger, than that induced by operators (4.13). It suffices to consider the case $A := r^{-1}\cos(\varphi)r\frac{\partial}{\partial r}$, which has no correspondence in (4.13); (clearly, φ is the variable on the unit circle $\mathbb{S}^1$).

The operator (4.14) is degenerate; i.e., the co-variable ρ belonging to r in its symbol is multiplied by r. This suggests studying such operators in r-direction in terms of the Mellin transform, while locally in x-direction we prefer the Fourier transform. Clearly, x is the variable on X. The Mellin transform on the half-axis $\mathbb{R}_+ \ni r$ is defined by

$$(Mu)(z) = \int_{\mathbb{R}_+} r^{-z} u(r) dr, \tag{4.15}$$

first as a continuous map

$$C_0^\infty(\mathbb{R}_+) \to \mathcal{A}(\mathbb{C})|_{\Gamma_{\frac{1}{2}}},$$

with $\mathcal{A}(\mathbb{C})$ being the space of entire functions in the complex z-plane. Here, we set

$$\Gamma_{\frac{1}{2}-\beta} := \{z \in \mathbb{C} : \mathrm{Re}(z) = \frac{1}{2} - \beta\} \tag{4.16}$$

for any $\beta \in \mathbb{R}$, where β has the meaning of a weight. This notation is motivated by the well-known fact, that the map $C_0^\infty(\mathbb{R}_+) \to \mathcal{A}(\mathbb{C})|_{\Gamma_{\frac{1}{2}-\beta}}$ extends by continuity to an isomorphism

$$M_\beta : L^{2,\beta}(\mathbb{R}_+) \to L^2(\Gamma_{\frac{1}{2}-\beta}). \tag{4.17}$$

Here, $L^{2,\beta}(\mathbb{R}_+) := r^\beta L^2(\mathbb{R}_+)$. Let us now pass to other weighted spaces, where the link $X \in \mathfrak{M}_0$ is compact and non-trivial, i.e. of dimension $n > 0$. First we look at spaces in local coordinates in $\mathbb{R}^n$. Sobolev spaces in $(r, x) \in \mathbb{R}_+ \times \mathbb{R}^n$ (first for the weight $\beta = 0$) will refer to a combination of the Mellin transform on $\mathbb{R}_+$ and the Fourier transform on $\mathbb{R}^n$. Then the space $\mathcal{H}^s(\mathbb{R}_+ \times \mathbb{R}^n)$ for any real s is defined to be the completion of $C_0^\infty(\mathbb{R}_+ \times \mathbb{R}^n)$ with respect to the norm

$$\|u\|_{\mathcal{H}^s(\mathbb{R}_+ \times \mathbb{R}^n)} := \left(\int_{\Gamma_{\frac{n+1}{2}}} \int_{\mathbb{R}^n} (1 + |z|^2 + |\xi|^2)|MFu(z, \xi)|^2 dz d\xi \right)^{\frac{1}{2}}. \tag{4.18}$$

Choosing an open covering $U_1, \ldots, U_N$ of X and a subordinate partition of unity $\varphi_1, \ldots, \varphi_N$, we can first form the spaces $\varphi_j \mathcal{H}^s(\mathbb{R}_+ \times U_j)$, where $\mathcal{H}^s(\mathbb{R}_+ \times U_j)$ is defined to be the pull-back of $\mathcal{H}^s(\mathbb{R}_+ \times \mathbb{R}^n)$ under the chart $U_j \to \mathbb{R}^n$, and then we set

$$\mathcal{H}^s(\mathbb{R}_+ \times X) := \sum_{j=1}^{N} \varphi_j \mathcal{H}^s(\mathbb{R}_+ \times U_j). \tag{4.19}$$

Moreover, we define the space

$$\mathcal{H}^{s,\beta}(\mathbb{R}_+ \times X) := r^\beta \mathcal{H}^s(\mathbb{R}_+ \times X) \tag{4.20}$$

for any real weight β. Note that we have $\mathcal{H}^{s,\beta}(\mathbb{R}_+ \times X) \subset H_{\text{loc}}^s(\mathbb{R}_+ \times X)$ for all s, β.

Our next objective is to introduce a slight modification of the spaces (4.20); the reason will be obvious below in connection with weighted edge spaces.

We choose a cut-off function $\omega(r)$ (i.e., a function in $C^\infty(\overline{\mathbb{R}}_+)$, which is $\equiv 1$ near $r = 0$ and vanishing for large r). Using the above notation we form the spaces $\mathcal{K}^{s,\beta}(\mathbb{R}_+ \times U_j) := \omega \mathcal{H}^{s,\beta}(\mathbb{R}_+ \times U_j) + (1 - \omega)H^s(\mathbb{R}_+ \times U_j)$ with $\mathcal{H}^{s,\beta}(\mathbb{R}_+ \times U_j)$ being the pull-back of $\mathcal{H}^{s,\beta}(\mathbb{R}_+ \times \mathbb{R}^n)$ under charts $\chi_j : U_j \to \mathbb{R}^n$; the space $H^s(\mathbb{R}_+ \times U_j) := H^s(\mathbb{R} \times U_j)|_{\mathbb{R}_+ \times U_j}$ is defined in a similar manner via pull-back of $H^s(\mathbb{R}_+ \times \mathbb{R}^n)$ under charts $\mathbb{R}_+ \times U_j \to \mathbb{R}_+ \times \mathbb{R}^n$ of the form $(r, x) \to (r, r\chi_j(x))$. Then we define

$$\mathcal{K}^{s,\beta}(\mathbb{R}_+ \times X) := \sum_{j=1}^{N} \varphi_j \mathcal{K}^{s,\beta}(\mathbb{R}_+ \times U_j). \tag{4.21}$$

Note, that $\mathbb{R}_+ \times X$ is interpreted as a (stretched) cone \ its tip, the link of which equals X. In several papers it is also denoted by $X^\wedge$. Similarly as (4.19) the latter definition is a non-direct sum of spaces in the corresponding topology.

Now consider a manifold B with edge $Y \in \mathfrak{M}_0$ of dimension q. By definition, we have $Y \subset B$ and $B \setminus Y$ is a C^∞-manifold. In addition, Y has a tubular (or collar) neighborhood V, which can be identified with some $X^\triangle$-bundle over Y, and there is an X-bundle V_1 over Y, with $Y \subset V_1 \subseteq V$. Moreover, there is an $\mathbb{R}_+$-action $\{\nu_\iota\}_{\iota \in \mathbb{R}_+}$ on V, such that $\nu_\iota \nu_{\iota'} = \nu_{\iota \iota'}$ for all $\iota, \iota' \in \mathbb{R}_+$, further ν_1 is the identity, and we have $\nu_\iota : V_1 \to V_1$ for $\iota \leq 1$ and $\nu_\iota : V \setminus V_1 \to V \setminus V_1$ for $\iota \geq 1$.

Similarly to Lemma 4.1 we form abstract edge spaces

$$\mathcal{W}^s(\mathbb{R}^q, E), \tag{4.22}$$

$s \in \mathbb{R}$, for a Hilbert space E, which is endowed with a group action $\{\kappa_\lambda\}_{\lambda \in \mathbb{R}_+}$, as the completion of $C_0^\infty(\mathbb{R}^q, E)$ with respect to the norm

$$\|u\|_{\mathcal{W}^s(\mathbb{R}^q, E)} := \left(\int [\eta]^{2s} \|\kappa^{-1}(\eta)(F_{y \to \eta} u)(\eta)\|_E^2 d\eta \right)^{1/2}, \tag{4.23}$$

where $\kappa(\eta) := \kappa_{[\eta]}$. Observe, that $\mathcal{S}(\mathbb{R}^q, E)$ is dense in $\mathcal{W}^s(\mathbb{R}^q, E)$, and we have (4.22) $\subset \mathcal{S}'(\mathbb{R}^q, E)$ for every $s \in \mathbb{R}$.

In the edge calculus this will be applied to $E := \mathcal{K}^{s,\beta}(\mathbb{R}_+ \times X)$ for $(\kappa_\lambda u)(r, x) := \lambda^{\frac{n+1}{2}} u(\lambda r, x)$. Remember, that before in Lemma 4.1 we considered the case $E = H^s(\mathbb{R}^{n+1})$ with

$$(\kappa_\lambda f)(r, x) := \lambda^{\frac{n+1}{2}} f(\lambda r, x), \tag{4.24}$$

where the origin $0 \in \mathbb{R}^{n+1}$ is regarded as a conical singularity with link $\mathbb{S}^n \ni x$, which is just the unit sphere in $\mathbb{R}^{n+1}$. Here we prefer the spaces $E := \mathcal{K}^{s,\beta}(\mathbb{R}_+ \times X)$ from Lemma 4.21 for the link $X \in \mathfrak{M}_0$, and $x \in X$, with an analogue of the above-mentioned group action, with $r \in \mathbb{R}_+$ being of similar meaning as before. Note, that

$$\mathcal{K}^{0,0}(\mathbb{R}_+ \times X) = \mathcal{H}^{0,0}(\mathbb{R}_+ \times X) = r^{-n/2} L^2(\mathbb{R}_+ \times X) \tag{4.25}$$

for $n = \dim(X)$, see also [74], Lemmas 4.3 and 4.6, (with L^2 referring to the measure $dr\,dx$), and the corresponding group action is unitary in the spaces (4.25). Because of

$$\mathcal{K}^{s,\beta}(\mathbb{R}_+ \times X) = \omega(r) \mathcal{H}^{s,\beta}(\mathbb{R}_+ \times X) + (1 - \omega(r)) H^s(\mathbb{R}_+ \times X) \tag{4.26}$$

the variable r on the half-axis is contained in both summands of $\mathcal{K}^{s,\beta}(\mathbb{R}_+ \times X)$. We also use the fact, that the intersection

$$\omega(r) \mathcal{H}^{s,\beta}(\mathbb{R}_+ \times X) \cap (1 - \omega(r)) H^s(\mathbb{R}_+ \times X)$$

is contained both in $\mathcal{H}^{s,\beta}(\mathbb{R}_+ \times X)$ and $H^s(\mathbb{R}_+ \times X)$. Therefore, the space $\mathcal{K}^{s,\beta}(\mathbb{R}_+ \times X)$ is independent of the choice of the cut-off function $\omega(r)$. Recall, that the function $u(r, x, y)$ in the norm expression (4.23) appears in the combination

$$\kappa^{-1}(\eta)(F_{y \to \eta} u)(r, x, \eta),$$

i.e., it has the form

$$[\eta]^{-\frac{n+1}{2}}(F_{y \to \eta} u)(r[\eta]^{-1}, x, \eta).$$

Globally, on a compact manifold B with edge Y we define the weighted edge Sobolev spaces

$$\mathcal{W}^{s,\beta}(B)$$

of smoothness $s \in \mathbb{R}$ and weight $\beta \in \mathbb{R}$ in terms of a covering of B by open sets

$$\{V_1, \ldots, V_N; U_1, \ldots, U_M\} \tag{4.27}$$

which together form an open covering of B, such that $V_j \cap Y =: Y_j \neq \emptyset$ for all $j = 1, \ldots, N$, and $U_k \cap Y = \emptyset$ for all $k = 1, \ldots, M$. Then $\{Y_1, \ldots, Y_N\}$ is an open covering of Y and $\{V_1 \setminus Y_1, \ldots, V_N \setminus Y_N, U_1, \ldots, U_M\}$ is an open covering of $B \setminus Y$. Let us define

$$\mathcal{W}^{s,\beta}(V_j) \cong \mathcal{W}^{s,\beta}(\mathbb{R}^q, \mathcal{K}^{s,\beta}(\mathbb{R}_+ \times X)) \tag{4.28}$$

belonging to an identification $V_j \to \mathbb{R}^q \times (\mathbb{R}_+ \times X)$, where the latter map contains a chart $Y_j \to \mathbb{R}^q$. Globally, the edge space $\mathcal{W}^{s,\beta}(B)$ of smoothness s and weight β is defined as a non-direct sum

$$\mathcal{W}^{s,\beta}(B) := \sum_{j=1}^{N} \psi_j \mathcal{W}^{s,\beta}(V_j) + \sum_{k=1}^{M} \varphi_k H^s_{\mathrm{loc}}(U_k), \tag{4.29}$$

with a chosen partition of unity $\{\psi_1, \ldots, \psi_N; \varphi_1, \ldots, \varphi_M\}$, subordinate to (4.27). Note that the summands containing φ_k are independent of β.

Lemma 4.4 *Let $B \in \mathfrak{M}_1$ be a compact manifold with edge Y. Then we have*

$$\mathcal{W}^{s,\beta}(B)|_{B \setminus Y} \subset H^s_{\mathrm{loc}}(B \setminus Y). \tag{4.30}$$

Proof Clearly it suffices to show, that $\varphi \mathcal{W}^{s,\beta}(B) \subset H^s_{\mathrm{loc}}(B \setminus Y)$ for every $\varphi \in C_0^\infty(B \setminus Y)$. However, Lemma 4.1 combined with the above localisation of spaces $\mathcal{K}^{s,\beta}(\mathbb{R}_+ \times X)$ far from $r = 0$, see (4.21), gives us the desired $H^s_{\mathrm{loc}}(B \setminus Y)$-property. In other words, relation (4.29) is justified, since

$$\psi_j \mathcal{W}^{s,\beta}(B)|_{V_j \cap U_k}$$

and

$$\varphi_k H^s_{\mathrm{loc}}(V_j \cap U_k)$$

coincide for all j, k. $\qquad\qquad\qquad\qquad\qquad\qquad\qquad\qquad\qquad\qquad\qquad\qquad\qquad\qquad\qquad\quad\square$

Lemma 4.4 shows, that the main attention of the edge calculus is focused on the analysis in the spaces $\mathcal{W}^{s,\beta}(\mathbb{R}^q, E)$ for $E = \mathcal{K}^{s,\beta}(\mathbb{R}_+ \times X)$, with E being equipped with the group action (4.24), at least what concerns weighted Sobolev spaces. This aspect will be deepened by adding a corresponding pseudo-differential calculus for edge-degenerate operators, see Sect. 4.3 below. However, the abstract pseudo-differential calculus between spaces $\mathcal{W}^{s,\beta}(\mathbb{R}^q, E)$ for general Hilbert spaces E with group action is interesting as a model for other applications. Therefore, we briefly sketch more elements of this calculus. This material will also be useful for the following Sect. 4.3.

The general group action $(\kappa_\lambda)_{\lambda \in \mathbb{R}_+}$ acting on a separable Hilbert space E is assumed to be strongly continuous with respect to $\lambda \in \mathbb{R}_+$. Furthermore, we assume that $\kappa_\lambda \kappa_\delta = \kappa_{\lambda\delta}$ and $\kappa_{\lambda^{-1}} = \kappa_\lambda^{-1}$ for all $\lambda, \delta \in \mathbb{R}_+$, and $\kappa_1 = \mathrm{id}_E$.

Proposition 4.5 *Given a Hilbert space E with group action $\{\kappa_\lambda\}_{\lambda \in \mathbb{R}_+}$, there are constants $C, M > 0$ satisfying*

$$\|\kappa_\lambda\|_{\mathcal{L}(E)} \leq C(\max\{\delta, \delta^{-1}\})^M. \tag{4.31}$$

A proof of this property was given before, see also [68], page 87.

Remark 4.6 Let E be a Hilbert space with group action satisfying (4.31). Then as an immediate consequence of (4.23) we have

$$H^{s+M}(\mathbb{R}^q, E) \subseteq \mathcal{W}^s(\mathbb{R}^q, E) \subseteq H^{s-M}(\mathbb{R}^q, E) \tag{4.32}$$

for every $s \in \mathbb{R}$.

Remark 4.7 Under the assumptions of Remark 4.6 we have

$$\mathcal{W}^s(\mathbb{R}^q, E) = \mathcal{W}^0(\mathbb{R}^q, E^s) \tag{4.33}$$

when E^s coincides with the space E, but endowed with the group action $\{\kappa_\lambda^s\}_{\lambda \in \mathbb{R}_+}$ defined by $\kappa_\lambda^s := \lambda^{-s}\kappa_\lambda, s \in \mathbb{R}$. Similarly we have $\mathcal{W}^{s+t}(\mathbb{R}^q, E) = \mathcal{W}^s(\mathbb{R}^q, E^{-t})$ for every $s, t \in \mathbb{R}$.

In addition, we have the obvious relation

$$\langle D_y \rangle^\mu := F_{\eta \to y}^{-1} \langle \eta \rangle^\mu F_{y' \to \eta} : \mathcal{W}^s(\mathbb{R}^q, E) \to \mathcal{W}^{s-\mu}(\mathbb{R}^q, E). \tag{4.34}$$

Remark 4.8 The spaces $\mathcal{W}^s(\mathbb{R}^q, E)$ for a Hilbert space E with scalar product $(e, f)_E$ and endowed with a group action

$$\{\kappa_\lambda\}_{\lambda \in \mathbb{R}_+}$$

have numerous other interesting properties, e.g.,

(i) $\mathcal{W}^s(\mathbb{R}^q, E)$ is a Hilbert space with the scalar product

$$\langle u, v \rangle_{\mathcal{W}^s(\mathbb{R}^q, E)} := \int_{\mathbb{R}^q} \langle \eta \rangle^{2s} (\kappa(\eta)^{-1} Fu(\eta), \kappa(\eta)^{-1} Fv(\eta))_E d\eta;$$

(ii) The Schwartz space $\mathcal{S}(\mathbb{R}^q, E)$ is densely embedded in $\mathcal{W}^s(\mathbb{R}^q, E)$;
(iii) The space of Schwartz distributions $\mathcal{S}'(\mathbb{R}^q, E)$ contains $\mathcal{W}^s(\mathbb{R}^q, E)$ for every s;

Theorem 4.9 *Let E be a Hilbert space, and $\{\kappa_\lambda\}_{\lambda \in \mathbb{R}_+}$ be a group action on E. Then the operator $\mathcal{M}_\varphi$ of multiplication by $\varphi \in \mathcal{S}(\mathbb{R}^q)$ induces a continuous operator*

$$\mathcal{M}_\varphi : \mathcal{W}^s(\mathbb{R}^q, E) \to \mathcal{W}^s(\mathbb{R}^q, E) \tag{4.35}$$

for every $s \in \mathbb{R}$. The mapping

$$\mathcal{S}(\mathbb{R}^q) \times \mathcal{W}^s(\mathbb{R}^q, E) \to \mathcal{W}^s(\mathbb{R}^q, E)$$

defined by $\langle \varphi, u \rangle \to \mathcal{M}_\varphi u$ is continuous for every s.

Proof For proving the continuity (4.35) it suffices to verify, that $\mathcal{M}_\varphi$ satisfies a corresponding $\mathcal{W}^s(\mathbb{R}^q, E)$-norm estimate on the dense subspace $\mathcal{S}(\mathbb{R}^q, E)$, see Remark 4.8, (ii). We use the fact, that the Fourier transform $F = F_{y \to \eta}$ induces an isomorphism $\mathcal{S}(\mathbb{R}^q, E) \to \mathcal{S}(\mathbb{R}^q, E)$. Moreover, $\varphi \in \mathcal{S}(\mathbb{R}^q)$ and $u \in \mathcal{S}(\mathbb{R}^q, E)$ implies $\varphi u \in \mathcal{S}(\mathbb{R}^q, E)$ and $F(\varphi u) = (F\varphi) * (Fu)$. Set

$$m(\eta) = \|[\eta]^s \kappa^{-1}(\eta) \int F\varphi(\eta - \xi) Fu(\xi) d\xi\|_E.$$

Then we have

$$\|\varphi u\|^2_{\mathcal{W}^s(\mathbb{R}^q, E)} = \int [\eta]^{2s} \|\kappa^{-1}(\eta) F(\varphi u)(\eta)\|^2_E d\eta = \|m\|^2_{L^2(\mathbb{R}^q)}.$$

We have $[\eta]^s \leq |c|^{|s|}[\eta - \xi]^{|s|}|\xi|^s$ for every $\xi, \eta \in \mathbb{R}^q$ and $s \in \mathbb{R}$, for a constant $c > 0$. Moreover, setting $K(\lambda) := \lambda$ for $\lambda \geq 1$, $K(\lambda) := \lambda^{-1}$ for $0 < \lambda \leq 1$, such that $\|\kappa_\lambda\|_E \leq K(\lambda)^M$, we have $K([\eta]/[\xi]) \leq c[\eta - \xi]$. Thus we obtain

$$m(\eta) \leq c_s \|\kappa^{-1}(\eta) \int [\eta - \xi]^{|s|} \hat{\varphi}(\eta - \xi)[\xi]^s \hat{u}(\xi) d\xi\|_E$$

which is equal to

$$c_s \|\int \kappa^{-1}_{[\eta]/[\xi]}[\eta - \xi]^{|s|-k}[\eta - \xi]^k \hat{\varphi}(\eta - \xi)[\xi]^s \kappa^{-1}(\xi)\hat{u}(\xi)d\xi\|_E$$

and

$$\leq \tilde{c}_s \int [\eta - \xi]^{|s|+M-k}[\eta - \xi]^k |\hat{\varphi}(\eta - \xi)|[\xi]^s \|\kappa^{-1}(\xi)\hat{u}(\xi)\|_E d\xi$$

which in turn is

$$\leq \tilde{c}_s C_k(\varphi) \int [\eta - \xi]^{|s|+M-k}[\xi]^s \|\kappa^{-1}(\xi)\hat{u}(\xi)\|_E d\xi$$

for $C_k(\varphi) = \sup_\xi [\xi]^k |\hat{\varphi}(\xi)|$. Let $g(\eta) = [\eta]^{|s|+M-k}$, $h(\eta) = [\eta]^s \|\kappa^{-1}(\eta)\hat{u}(\eta)\|_E$, and chose $k \in \mathbb{N}$ so large that $|s| + M - k < -q$. Then we have $g \in L^1(\mathbb{R}^q)$, $h \in L^2(\mathbb{R}^q)$ and $\|h\|_{L^2(\mathbb{R}^q)} = \|u\|_{\mathcal{W}^s(\mathbb{R}^q, E)}$. Then the above inequality may be written in the form

$$m(\eta) \leq \tilde{c}_s C_k(\varphi)(g * h)(\eta),$$

$\eta \in \mathbb{R}^q$. Applying Young's inequality we obtain

$$\|\varphi u\|_{\mathcal{W}^s(\mathbb{R}^q, E)} = \|m\|_{L^2(\mathbb{R}^q)} \leq \tilde{c}_s C_k(\varphi)\|(g * h)\|_{L^2(\mathbb{R}^q)},$$

and this is

$$\leq \tilde{c}_s C_k(\varphi)\|g\|_{L^1(\mathbb{R}^q)}\|h\|_{L^2(\mathbb{R}^q)} = C_s C_k(\varphi)\|u\|_{\mathcal{W}^s(\mathbb{R}^q, E)}.$$

We see that $\mathcal{M}_\varphi$ is continuous as well as $(\varphi, u) \to \varphi u$. $\qquad \square$

Definition 4.10 For any open $\Omega \subseteq \mathbb{R}^q$ we define

$$\mathcal{W}^s_{\mathrm{comp}}(\Omega, E) := \{u \in \mathcal{W}^s(\mathbb{R}^q, E) : \mathrm{supp}(u) \Subset \Omega\};$$

Here, $\Subset \Omega$ indicates a compact subset of Ω.

$$\mathcal{W}^s_{\mathrm{loc}}(\Omega, E) := \{u \in \mathcal{D}'(\Omega, E) : \varphi u \in \mathcal{W}^s_{\mathrm{comp}}(\Omega, E)\}$$

for every $\varphi \in C_0^\infty(\Omega)$.

4.3   Operators in Edge Spaces

The pseudo-differential analysis on manifolds with edge refers to a large extent to operator-valued symbols, according to the following definition.

Definition 4.11 Let E and $\tilde{E}$ be Hilbert spaces, endowed with group actions $(\kappa_\lambda)_{\lambda \in \mathbb{R}_+}$ and $(\tilde{\kappa}_\lambda)_{\lambda \in \mathbb{R}_+}$, respectively. Then

$$S^\mu(\Omega \times \mathbb{R}^q; E, \tilde{E}), \tag{4.36}$$

for open $\Omega \subseteq \mathbb{R}^q$, is defined to be the space of all $a(y, \eta) \in C^\infty(\Omega \times \mathbb{R}^q, \mathcal{L}(E, \tilde{E}))$, satisfying the symbolic estimates

$$\|\tilde{\kappa}^{-1}(\eta)\{D_y^\alpha D_\eta^\beta a(y, \eta)\}\kappa(\eta)\|_{\mathcal{L}(E, \tilde{E})} \le c\langle\eta\rangle^{\mu-\beta} \tag{4.37}$$

for all $\alpha, \beta \in \mathbb{N}^q$ and all $y \in K$ for compact $K \subset \Omega$ and all $\eta \in \mathbb{R}^q$, with constants $c = c(\alpha, \beta, K) > 0$.

The space $S^\mu(\Omega \times \mathbb{R}^q; E, \tilde{E})$ is Fréchet with the semi-norm system

$$a \to \sup_{y \in K, \eta \in \mathbb{R}^q} \langle\eta\rangle^{-\mu+|\beta|}\|\tilde{\kappa}^{-1}(\eta)\{D_y^\alpha D_\eta^\beta a(y, \eta)\}\kappa(\eta)\|_{\mathcal{L}(E, \tilde{E})}$$

for compact $K \subset \Omega$ and $\alpha, \beta \in \mathbb{N}^q$.

Definition 4.12 By

$$S^{(\mu)}(\Omega \times \mathbb{R}^q; E, \tilde{E}) \tag{4.38}$$

for $\mu \in \mathbb{R}$ and $E, \tilde{E}$ as in Definition 4.11 we denote the space of all $a^{(\mu)}(y, \eta) \in C^\infty(\Omega \times (\mathbb{R}^q \setminus 0), \mathcal{L}(E, \tilde{E}))$ with

$$a^{(\mu)}(y, \lambda\eta) = \lambda^\mu \tilde{\kappa}_\lambda a(y, \eta)\kappa_\lambda^{-1} \tag{4.39}$$

for all $\lambda \in \mathbb{R}_+$. Moreover, by

$$S_{\mathrm{cl}}^{(\mu)}(\Omega \times \mathbb{R}^q; E, \tilde{E}) \tag{4.40}$$

we denote the subspace of all $a(y, \eta) \in S^\mu(\Omega \times \mathbb{R}^q; E, \tilde{E})$, such that

$$a(y, \eta) - \chi(\eta) \sum_{j=0}^{N} a^{(\mu-j)}(y, \eta) \in S^{\mu-(N+1)}(\Omega \times \mathbb{R}^q; E, \tilde{E}),$$

for every $N \in \mathbb{N}$, where $\chi(\eta)$ is any excision function. Finally, we set

$$S^{-\infty}(\Omega \times \mathbb{R}^q; E, \tilde{E}) := \bigcap_{N \in \mathbb{N}} S^{\mu-N}(\Omega \times \mathbb{R}^q; E, \tilde{E}) \tag{4.41}$$

We write $S_{(\mathrm{cl})}^{\mu}(\Omega \times \mathbb{R}^q; E, \tilde{E})$ if a statement is true both for $S^\mu(\Omega \times \mathbb{R}^q; E, \tilde{E})$ and $S_{\mathrm{cl}}^{\mu}(\Omega \times \mathbb{R}^q; E, \tilde{E})$.

Note, that

$$S_{(\mathrm{cl})}^{\mu}(\Omega \times \mathbb{R}^q; E, \tilde{E}) = C^\infty(\Omega, S_{(\mathrm{cl})}^{\mu}(\mathbb{R}^q; E, \tilde{E})) = C^\infty(\Omega) \hat{\otimes}_\pi S_{(\mathrm{cl})}^{\mu}(\mathbb{R}^q; E, \tilde{E}))$$

with $S_{(\mathrm{cl})}^{\mu}(\mathbb{R}^q; E, \tilde{E}))$ being the respective space of symbols, that are independent of y.

According to a well-known lemma on projective tensor products of elements in locally convex spaces every element $a(y, \eta) \in S_{(\mathrm{cl})}^{\mu}(\Omega \times \mathbb{R}^q; E, \tilde{E})$ can be written as a convergent sum

$$a(y, \eta) = \sum_{j=1}^{\infty} \lambda_j f_j \otimes \tilde{f}_j \tag{4.42}$$

for $\lambda_j \in \mathbb{C}$ with $\sum |\lambda_j| < \infty$ and sequences $\{f_j\}, \{\tilde{f}_j\}$ in $C^\infty(\Omega)$ and $S_{(\mathrm{cl})}^{\mu}(\mathbb{R}^q; E, \tilde{E}))$, respectively, tending to zero in the corresponding spaces.

Proposition 4.13 *Let*

$$a(\eta) \in S_{(\mathrm{cl})}^{\mu}(\mathbb{R}^q; E, \tilde{E}) \tag{4.43}$$

Then $\mathrm{op}(a)$ *induces a continuous operator*

$$\mathrm{op}(a) : \mathcal{W}^s(\mathbb{R}^q, E) \to \mathcal{W}^{s-\mu}(\mathbb{R}^q, \tilde{E}), \tag{4.44}$$

for every $s \in \mathbb{R}$, *and we have*

$$\|\mathrm{op}(a)\|_{\mathcal{L}(\mathcal{W}^s(\mathbb{R}^q, E), \mathcal{W}^{s-\mu}(\mathbb{R}^q, \tilde{E}))} \leq c[\eta]^{-\mu} \sup_{\eta \in \mathbb{R}^q} \left\| \tilde{\kappa}^{-1}(\eta) a(\eta) \kappa(\eta) \right\|_{\mathcal{L}(E, \tilde{E})} \tag{4.45}$$

for some constant $c > 0$.

Proof For $\hat{u}(\eta) = F_{y \to \eta} u$ we have

$$\|\mathrm{op}(a) u\|^2_{\mathcal{W}^{s-\mu}(\mathbb{R}^q, \tilde{E})} = \int [\eta]^{2(s-\mu)} \left\| \tilde{\kappa}^{-1}(\eta) a(\eta) \hat{u}(\eta) \right\|^2_{\tilde{E}} d\eta \tag{4.46}$$

which is equal to

$$\int [\eta]^{2(s-\mu)} \left\| [\tilde{\kappa}^{-1}(\eta) a(\eta) \kappa(\eta)][\kappa(\eta)^{-1} \hat{u}(\eta)] \right\|^2_{\tilde{E}} d\eta. \tag{4.47}$$

The latter expression is

$$\leq \int [\eta]^{2(s-\mu)} \left\| [\tilde{\kappa}^{-1}(\eta) a(\eta) \kappa(\eta) \right\|^2_{\mathcal{L}(E, \tilde{E})} \left\| \kappa(\eta)^{-1} \hat{u}(\eta) \right\|^2_{E} d\eta. \tag{4.48}$$

This in turn can be estimated by

$$c \sup_{\eta \in \mathbb{R}^q} [\eta]^{-2\mu} \left\| [\tilde{\kappa}^{-1}(\eta) a(\eta) \kappa(\eta) \right\|^2_{\mathcal{L}(E, \tilde{E})} \|u\|^2_{\mathcal{W}^s(\mathbb{R}^q, E)} , \tag{4.49}$$

which completes the proof. $\qquad\square$

We now extend Proposition 4.13 to the case of symbols with variable coefficients.

Proposition 4.14 *Let*

$$a(y, \eta) \in \mathcal{S}(\mathbb{R}^q_y) \hat{\otimes}_\pi S^\mu_{(\mathrm{cl})}(\mathbb{R}^q_\eta; E, \tilde{E}). \tag{4.50}$$

Then $\mathrm{op}(a)$ *induces a continuous operator*

$$\mathrm{op}(a) : \mathcal{W}^s(\mathbb{R}^q, E) \to \mathcal{W}^{s-\mu}(\mathbb{R}^q, \tilde{E}). \tag{4.51}$$

for every $s \in \mathbb{R}$,

Proof Applying relation (4.42) to $a(y, \eta) \in \mathcal{S}(\mathbb{R}^q_y) \hat{\otimes}_\pi S^\mu_{(\mathrm{cl})}(\mathbb{R}^q_\eta; E, \tilde{E})$ which is the same as $\mathcal{S}(\mathbb{R}^q_y, S^\mu_{(\mathrm{cl})}(\mathbb{R}^q_\eta; E, \tilde{E}))$, the symbol can be written in the form

$$a(y, \eta) = \sum_{j=1}^{\infty} \lambda_j \varphi_j(y) a_j(\eta)$$

for elements $\lambda_j \in \mathbb{C}$ with $\sum_{j=1}^{\infty} \lambda_j < \infty$ and $\varphi_j(y) \in \mathcal{S}(\mathbb{R}^q)$ and $a_j(\eta) \in S_{(\mathrm{cl})}^{\mu}(\mathbb{R}_\eta^q; E, \tilde{E})$, tending to 0 in the respective spaces. It suffices now to prove, that

$$\mathrm{op}(a) = \sum_{j=1}^{\infty} \lambda_j \varphi_j \mathrm{op}(a_j)$$

converges in $\mathcal{L}(\mathcal{W}^s(\mathbb{R}^q, E), \mathcal{W}^{s-\mu}(\mathbb{R}^q, \tilde{E}))$. However, this is an obvious consequence of Theorem 4.9 and Proposition 4.13. $\qquad\square$

It is useful to look at a special case of Proposition 4.13 to Sobolev spaces $H^s(\mathbb{R}^{q+m}) = \mathcal{W}^s(\mathbb{R}^q, E)$, for $E := H^s(\mathbb{R}^m))$ and $H^{s-\mu}(\mathbb{R}^{q+m}) = \mathcal{W}^{s-\mu}(\mathbb{R}^q, \tilde{E})$ for $\tilde{E} := H^{s-\mu}(\mathbb{R}^m))$, with E and $\tilde{E}$ being endowed with the group action, suggested by Lemma 4.1 for the case $m = n + 1$. Consider a scalar symbol $p(\eta, \tilde{\xi}) \in S^\mu(\mathbb{R}^{q+m}_{(\eta, \tilde{\xi})})$ with constant coefficients. We want to construct an operator-valued symbol $a(\eta) \in S^\mu(\mathbb{R}^q; E, \tilde{E})$, such that

$$\mathrm{op}_{(y, \tilde{x})}(p) : H^s(\mathbb{R}^{q+m}) \to H^{s-\mu}(\mathbb{R}^{q+m}) \tag{4.52}$$

can be rephrased by

$$\mathrm{op}_y(a) : \mathcal{W}^s(\mathbb{R}^q, E) \to \mathcal{W}^{s-\mu}(\mathbb{R}^q, \tilde{E}). \tag{4.53}$$

Using the isomorphism

$$\mathfrak{U} : H^s(\mathbb{R}^{q+m}) \to \mathcal{W}^s(\mathbb{R}^q, H^s(\mathbb{R}^m))$$

we obtain the relation

$$\mathrm{op}_y(a) = \mathfrak{U}^{-1} \circ \mathrm{op}_{(y, \tilde{x})}(p) \circ \mathfrak{U}. \tag{4.54}$$

This corresponds to the identity

$$a(\eta) = \mathrm{op}_{\tilde{x}}(p(\frac{\tilde{\xi}}{[\eta]}, \eta)). \tag{4.55}$$

4.4   Notes on Pseudo-Differential Boundary Problems

In order to motivate structures on manifolds with edge we briefly comment aspects of a special case, concerning elliptic boundary problems in the spirit of L. Boutet de Monvel [9] based on symbols with the transmission property at the boundary. We drop here relations to the Atiyah-Singer theorem and to other well-known achievements of numerous authors, such as L. Hörmander [34], G.I. Eskin [24], G. Grubb [30].

In order to fix notation, let us briefly recall some basics on ellipticity of pseudo-differential operators on closed compact C^∞-manifolds $X \in \mathfrak{M}_0$. Let $\mathrm{Vect}(X)$ denote the set of all complex smooth vector bundles over X. This is a semi-group under direct sum $\oplus$, and it can be completed to a commutative group, namely, $K(X)$ with the operation $\oplus$, which is just the background of well-known considerations on the index-Theorem of Atiyah-Singer. For later purposes we outline here some aspects of this material. Elliptic pseudo-differential symbols (here for simplicity classical) $a(x, \xi)$ on X of order $\mu \in \mathbb{R}$ are modulo lower order terms determined by the invertibility of its homogeneous principal symbol $a^{(\mu)}(x, \xi)$ of order μ, defined for $\xi \neq 0$, taking values in isomorphisms $p^*(J^-) \to p^*(J^+)$ for elements $J^\pm \in \mathrm{Vect}(X)$, with p^* being the pull-back under the canonical projection $p : T^*(X) \setminus 0 \to X$. Operators A determined by an elliptic symbol a are called elliptic. Those A are known to be Fredholm, both as maps

$$A : H^s(X, J^-) \to H^{s-\mu}(X, J^+) \tag{4.56}$$

with $H^s(X, J), s \in \mathbb{R}$, denoting Sobolv spaces of distributional sections in $J \in \mathrm{Vect}(X)$, and

$$A : C^\infty(X, J^-) \to C^\infty(X, J^+), \tag{4.57}$$

with $C^\infty(X, J)$ being the space of smooth sections in the bundle J, where the index of (4.56) and (4.57) are the same. Recall, that a bounded operator A is said to be Fredholm, if $\mathrm{ind}(A) = \dim(\ker(A)) - \dim(\mathrm{coker}(A))$ is finite. Below we shall employ the fact, that we have

Theorem 4.15 *For every μ and any J there exists an elliptic operator A which is bijective between the respective spaces, for $J = J^+ = J^-$.*

Proof It suffices to take an operator $A(\lambda)$ which is parameter-dependent elliptic, with parameter $\lambda \in \mathbb{R}$, and to fix the parameter sufficiently large. $\qquad\square$

We will use the simple fact, that Fredholm operators $A : H \to \tilde{H}$ (say, between Hilbert spaces $H, \tilde{H}$) may be considered in connection with isomorphisms

$$
\mathbb{A} = \begin{pmatrix} A & k \\ t & q \end{pmatrix} : \begin{matrix} H \\ \oplus \\ \mathbb{C}^{N_-} \end{matrix} \to \begin{matrix} \tilde{H} \\ \oplus \\ \mathbb{C}^{N_+} \end{matrix} , \tag{4.58}
$$

for certain N_-, N_+ with $\mathrm{ind}(A) = N_- - N_+$ and finite-dimensional operators k, t, q, depending on A.

We now pass to linear elliptic boundary value problems in the sense of L. Boutet de Monvel [9]. Those are formally represented by 2×2 block matrices of operators

$$
\mathcal{A} = \begin{pmatrix} A + G & K \\ T & Q \end{pmatrix} : \begin{matrix} H_1 \\ \oplus \\ H_1' \end{matrix} \to \begin{matrix} H_2 \\ \oplus \\ H_2' \end{matrix} , \tag{4.59}
$$

where $A : H_1 \to H_2$ is a pseudo-differential operator on a manifold X with smooth boundary ∂X with the transmission property at ∂X and $G : H_1 \to H_2$ is a so-called Green operator (see details below in this subsection). Moreover, $T : H_1 \to H_2'$ is a trace operator, representing boundary conditions, $K : H_1' \to H_2$ is a co-boundary condition (in the terminology of some authors, dual in some sense to a trace operator), here called a potential operator, and $Q : H_1' \to H_2'$ is a pseudo-differential operator on the boundary. The action of operators in (4.59), may concern L^2-based Sobolev spaces on X and ∂X, respectively, of some smoothness, shifted by pseudo-differential orders involved in (4.59). Classical examples are boundary value problems for Laplacians, e.g., the Dirichlet or the Neumann problem. Both have the form of column matrices

$$
\mathcal{A} = \begin{pmatrix} A \\ T \end{pmatrix} : H_1 \to \begin{matrix} H_2 \\ \oplus \\ H_2' \end{matrix} , \tag{4.60}
$$

where A is the Laplacian Δ itself, say, when X is a smooth bounded domain in $\mathbb{R}^n$. In the case of the Dirichlet problem, T is the operator of restriction to ∂X. We shall outline here some well-known facts, to see some similarities of boundary value problems with the edge calculus, but also inevitable differences. It is well-known, that we have different choices for the spaces in (4.60). We may take, e.g., Sobolev spaces $H_1 := H^s(X, E_1)$, $H_2 := H^{s-2}(X, E_2)$, with E_1, E_2 being elements of $\mathrm{Vect}(X)$ (which are, for simplicity, first assumed to be equal $X \times \mathbb{C}$, and thus $H_1 = H^s(X)$, $H_2 = H^{s-2}(X)$) and T the restriction operator $H^s(X) \to H_2' := H^{s-1/2}(\partial X)$ for $s > 0$ sufficiently large. For the moment we do not comment here other natural choices of spaces in (4.60). In the special case of the Dirichket problem the operator (4.60) is even an isomorphism with the inverse

$$\mathcal{P} = \left(P + G \; K \right) : \begin{matrix} H_2 \\ \oplus \\ H_2' \end{matrix} \to H_1. \tag{4.61}$$

In classical potential theory $P + G$ is called Green's function of $\Omega := X \setminus \partial X$ and K the Poisson kernel, related to the Dirichlet problem. At the same time, (4.61) is a parametrix of (4.60), i.e., in the compositions $\mathcal{A}\mathcal{P} = \mathcal{I}_2 + \mathcal{C}_2$ and $\mathcal{P}\mathcal{A} = \mathcal{I}_1 + \mathcal{C}_1$ we have the corresponding identity maps $\mathcal{I}_1, \mathcal{I}_2$ and compact operators $\mathcal{C}_1, \mathcal{C}_2$ in the respective spaces (vanishing in this special case). More generally, ellipticity of $\mathcal{A}$ like (4.59) in Boutet de Monvel's algebra entails the existence of a parametrix

$$\mathcal{P} = \begin{pmatrix} P + G_1 & K_1 \\ \\ R & S \end{pmatrix} : \begin{matrix} H_2 \\ \oplus \\ H_2' \end{matrix} \to \begin{matrix} H_1 \\ \oplus \\ H_1' \end{matrix}, \tag{4.62}$$

which is of a similar meaning as before. It remains to define ellipticity in the context of the underlying symbolic structure. We do not recall here all aspects of the involved symbols. It suffices to know for the present discussion, that the pseudo-differential operator $A \in L^{\mu}_{\mathrm{cl}}(X \setminus \partial X)_{\mathrm{trans}}$ has the above-mentioned transmission property with respect to ∂X (for simplicity, concerning A we assume here classical symbols and scalar operators in the upper left corners), and for the transmission property we refer to expositions of the authors, mentioned at the beginning of this subsection, or to the monographs [61], see also [44] Chapter 2, or [64].

In order to deepen further details, let us note, that the achievements around ellipticity of pseudo-differential operators on smooth manifolds, including boundary problems, became a source of new ideas, initiated by the work of S. Agmon, A. Douglis, and L. Nirenberg [1], I.M. Gelfand [26], F. Hirzebruch [32], M.F. Atiyah and I.M. Singer [5], M.F. Atiyah and R. Bott [3], L. Boutet de Monvel [9], and many others. The present "keywords" about symbols of operators and K-theoretic background on the index of associated Fredholm operators may illustrate the wealth of structures belonging to the future of singular analysis in the context of ellipticity. We completely leave out here similar perspectives, concerning parabolicity, and other related topics, see, e.g., the work of T. Krainer [41], and the bibliography there.

Symbols of elliptic operators near the boundary of a smooth manifold X with boundary ∂X, more precisely, in a collar neighbohood of ∂X with local coordinates $x = (x', x_n)$ in $\overline{\mathbb{R}}^n_+$ for variables $x' := (x_1, \ldots, x_{n-1})$ locally tangent to the boundary, and $x_n = 0$ along the boundary, where X itself is described by $x_n \geq 0$, will be written in the form $a(x, \xi) := a(x', x_n, \xi', \xi_n)$ with $\xi = (\xi', \xi_n) \in \mathbb{R}^{n-1} \times \mathbb{R}$ being the corresponding co-variables. In simplest cases they are assumed to be of integer order μ, belong to

$$S^{\mu}_{\mathrm{cl}}(\mathbb{R}^n_x \times \mathbb{R}^n_{\xi})|_{x_n \geq 0} := S^{\mu}_{\mathrm{cl}}(\overline{\mathbb{R}}^n_+ \times \mathbb{R}^n),$$

and have the so-called transmission property with respect to $x_n = 0$, indicated by $S^\mu_{\text{cl}}(\overline{\mathbb{R}}_+ \times \mathbb{R}^n)_{\text{trans}}$.

The notation "transmission property" is used by several authors in different meaning. For simplicity we give here the most simple variant, in order to illustrate, to what extent it is a specific property of symbols near the boundary, locally close to $x_n = 0$.

Definition 4.16 An element $a(x', x_n, \xi', \xi_n) \in S^\mu_{\text{cl}}(\overline{\mathbb{R}}^n_+ \times \mathbb{R}^n)$ for $\mu \in \mathbb{Z}$ is said to have the transmission property at $x_n = 0$, i.e., to belong to $S^\mu_{\text{cl}}(\overline{\mathbb{R}}^n_+ \times \mathbb{R}^n)_{\text{trans}}$, if the homogeneous components $a^{(\mu-j)}$ (of order $\mu - j$ in $\xi \neq 0$) of a satisfy the condition

$$(D^k_{x_n} D^\alpha_\xi a^{(\mu-j)})(x', 0, 0, 1) = (-1)^{\mu-j-|\alpha|}(D^k_{x_n} D^\alpha_\xi a^{(\mu-j)})(x', 0, 0, -1) \qquad (4.63)$$

for all $x' \in \mathbb{R}^{n-1}$ and $j, k \in \mathbb{N}, \alpha \in \mathbb{N}^n$.

Clearly, "0, 0" in expression (4.63) means $x_n = 0, \xi = 0$. Pseudo-differential operators with symbols in $S^\mu_{\text{cl}}(\overline{\mathbb{R}}^n_+ \times \mathbb{R}^n)_{\text{trans}}$ are also said to have the transmission property at the boundary. This property also makes sense globally on a manifold with smooth boundary, because of the coordinate invariance under diffeomorphisms, which are smooth up to the boundary. It is well-known, that any such a induces a continuous operator

$$\text{op}^+(a) := \text{e}^+ \text{op}(a)\text{r}^+ : C^\infty_0(\overline{\mathbb{R}}^n_+) \to C^\infty(\overline{\mathbb{R}}^n_+), \qquad (4.64)$$

with $\text{r}^+ : C^\infty_0(\overline{\mathbb{R}}^n_+) \to \mathcal{E}'(\mathbb{R}^n)$ being defined as the operator of extending elements by zero to $\mathbb{R}^n$ for $x_n < 0$ (where $\mathcal{E}'$ is the space of compactly supported Schwartz distributions on the respective open set), and e^+ is the operator of restriction to $\mathbb{R}^n_+$. The mapping property (4.64) is just a consequence of the transmission property. A known result in this context is, that (4.64) extends to a continuous operator

$$\text{op}^+(a) : H^s(\mathbb{R}^n_+) \to H^{s-\mu}(\mathbb{R}^n_+), \qquad (4.65)$$

here employed for sufficiently large $s \in \mathbb{R}, s > 0$, where $H^s(\mathbb{R}^n_+) := H^s(\mathbb{R}^n)|_{\mathbb{R}^n_+}$. While the correspondence $a \mapsto \text{op}(a)$ is an analogue of a procedure in quantum physics (where a Hamilton function on a phase-space is related to a Hamilton operator, e.g., the impulse variable ξ turns to $i^{-1}\partial/\partial x$), the correspondence $a \mapsto \text{op}^+(a)$ is sometimes called "truncation quantization", see also a paper jointly with J. Seiler [72]. Examples of symbols with the transmission property at the boundary are polynomials in ξ. Those are associated with differential operators. By the way, symbols of differential operators also have the so-called minus-property. This includes the condition, that such a symbol admits an analytic extension with respect to the covariable $\zeta := \xi + i\tau \in \mathbb{C}$ for $\tau \geq 0$, where the real part is just the covariable ξ, we are talking about. Then smoothness under the action of $\text{op}^+(a)$ up to the boundary is also preserved. Clearly, this notation depends on a sign-convention in the definition of the Fourier transform, see also Eskin's monograph [24], where the opposite sign is used. Furthermore, compositions and Leibniz-inverses (in case of ellipticity) preserve this property. As noted before, truncation makes sense also on a (say compact) smooth manifold X with boundary; then relations (4.64) and (4.65) have corresponding global analogues, namely,

$$\mathrm{op}^+(a) : C^\infty(X) \to C^\infty(X), H^s(X) \to H^{s-\mu}(X), \tag{4.66}$$

Operators $A = \mathrm{op}^+(a)$ for elliptic symbols a on a smooth manifold with boundary in general do not induce Fredholm operators between the involved spaces, see (4.65). Ellipticity in this context is determined by the homogeneous principal symbol

$$\sigma_X^{(\mu)}(A)(x, \xi) := a^{(\mu)}(x, \xi), \tag{4.67}$$

which is invariantly defined as a complex-valued homogeneous function on $T^*(X) \setminus 0$, of homogeneity μ. For convenience we interpret (x, ξ) as a global variable, though it always refers to local charts on the underlying space.

An example is the Laplacian Δ in a bounded smooth domain $\Omega \subset \mathbb{R}^n$; in this case we have $X = \overline{\Omega}$. We intend to illustrate results for elliptic differential operators and associated elliptic boundary condition. First of all, the maps (4.66) are surjective, since the involved symbols are minus-functions. Thus. in order to associate a Fredholm map with A, we have to pass to a column matrix like (4.60), containing a boundary condition T, by notation a trace operator in the terminology of [9]. For the Laplacian it suffices to take Dirichlet conditions

$$T := \mathrm{r}' : H^s(X) \to H^{s-1/2}(\partial X),$$

which induces a surjective map $\mathrm{r}' : \ker(A) \to H^{s-1/2}(\partial X)$ for $A = \Delta$, where r' is just the restriction operator to the boundary. The corresponding operator (4.60) is Fredholm, which is a special example of an elliptic boundary problem, satisfying the Shapiro-Loptinski-condition, represented by the operator T. Other examples may be of the form $T = \mathrm{r}'D$ for some differential operator D on $\Omega = X \setminus \partial X$ with smooth coefficients up to the boundary. If D is of order ν, then $T : H^s(X) \to H^{s-\nu-1/2}(\partial X)$ is continuous. Locally close to the boundary we ask, that the so-called boundary symbol of (4.60), denoted by

$$\sigma_{\partial X}^{(\mu)}\left(\begin{pmatrix} A \\ T \end{pmatrix} \right)(x', \xi') \tag{4.68}$$

is (locally) defined as

$$
\begin{pmatrix} \mathrm{r}^+ \mathrm{op}_{x_n}(a^{(\mu)}|_{x_n=0})\mathrm{e}^+ \\[2ex] \mathrm{r}' t(D_x) \end{pmatrix} (x', \xi') : H^s(\mathbb{R}_{x_n,+}) \to \begin{matrix} H^{s-\mu}(\mathbb{R}_{x_n,+}) \\ \oplus \\ \mathbb{C}^m \end{matrix} \tag{4.69}
$$

for some polynomial $t(D_x)$ in $D_{x_1}, \ldots, D_{x_n}$, is a bijective operator function on $T^*(\partial X \setminus X)$, with X representing the zero-section in $T^*(\partial X)$. Here we only consider trace operators T in integral form, (called to be of class zero in [9]). Those are assumed to be of the form $T = \mathrm{op}_{x'}(t)$ with an operator-valued symbol $t(x', \xi') \in S^\mu_{\mathrm{cl}}(\mathbb{R}^{n-1}_{x'} \times \mathbb{R}^{n-1}_{\xi'}; \mathcal{S}(\overline{\mathbb{R}}_+), \mathbb{C}^m)$ with homogeneous principal symbol $\sigma^{(\mu)}_{\partial X}(\mathcal{A})(x', \xi') := t^{(\mu)}(x', \xi') \in S^{(\mu)}(\mathbb{R}^{n-1}_{x'} \times \mathbb{R}^{n-1}_{\xi'} \setminus 0; \mathcal{S}(\overline{\mathbb{R}}_+), \mathbb{C}^m)$. Here (x', ξ') is also interpreted as a global variable in $T^*(\partial X \setminus 0)$. In sum, the couple of symbols

$$
\{\sigma^{(\mu)}_X(\mathcal{A})(x, \xi), \sigma^{(\mu)}_{\partial X}(\mathcal{A})(x', \xi')\} \tag{4.70}
$$

for $\sigma^{(\mu)}_X(\mathcal{A})(x, \xi) = \sigma^{(\mu)}_X(A)(x, \xi)) = a^{(\mu)}(x, \xi)$, $\mathcal{A} := \begin{pmatrix} A \\ T \end{pmatrix}$. Then the ellipticity of $\mathcal{A}$ is just defined as the bijectivity of both components of (4.70). It remains to extend the definition to block-matrices (4.59), i.e., to include the entries K and Q. The operator Q is simply a (system of) classical pseudo-differential operators on ∂X of order μ. The potential operator K is a pseudo-differential map $K := \mathrm{op}_{x'}(k)$ for an operator-valued symbol $k(x', \xi') \in S^\mu_{\mathrm{cl}}(\mathbb{R}^{n-1}_{x'} \times \mathbb{R}^{n-1}_{\xi'}; \mathbb{C}^N, \mathcal{S}(\overline{\mathbb{R}}_+))$.

Let us now extend this consideration to the case that

$$
\sigma^{(\mu)}_{\partial X}(A)(x', \xi') : H^s(\mathbb{R}_{x_n,+}) \to H^{s-\mu}(\mathbb{R}_{x_n,+}) \tag{4.71}
$$

is not surjective. Since the ellipticity of A implies, that (4.71) is a family of Fredholm operators, we find an $N \in \mathbb{N}$, and an operator family $(\sigma_{\partial X}(K))(x', \xi') : \mathbb{C}^N \to \mathcal{S}(\overline{\mathbb{R}}_{x_n,+})$ with $k^{(\mu)}(x', \xi') \in S^\mu(\mathbb{R}^{n-1}_{x'} \times \mathbb{R}^{n-1}_{\xi'} \setminus 0; \mathbb{C}^N, \mathcal{S}(\overline{\mathbb{R}}_{x_n,+}))$ such that

$$
\left(\sigma^{(\mu)}_{\partial X}(A), \sigma_{\partial X}(K)\right)(x', \xi') : \begin{pmatrix} H^s(\mathbb{R}_{x_n,+}) \\ \oplus \\ \mathbb{C}^N \end{pmatrix} \to H^{s-\mu}(\mathbb{R}_{x_n,+}) \tag{4.72}
$$

is surjective. The family of kernels of over $T^*(\partial X \setminus 0)$ (with 0 representing the zero section of $T^*(\partial X)$), where we assume, for simplicity, that it is the pull-back $\pi^*_{\partial X}(J^-)$ of some other bundle J^- over ∂X under the canonical projection $T^*(\partial X \setminus 0) \to \partial X$. Then we can add to (4.72) a row of mappings

$$\left(\sigma_{\partial X}^{(\mu)}(T), \sigma_{\partial X}(Q)\right)(x', \xi') : \begin{array}{c} H^s(\mathbb{R}_{x_n,+}) \\ \oplus \\ \mathbb{C}^N \end{array} \to \pi_{\partial X}^*(J^+) \qquad (4.73)$$

for another vector bundle J^+, satisfying analogous conditions as before for J^-, which gives rise altogether to a family of block-matrix isomorphisms

$$\sigma_{\partial X}^{(\mu)}(\mathcal{A})(x', \xi') = \begin{pmatrix} \sigma_{\partial X}^{(\mu)}(A) & \sigma_{\partial X}(T) \\ \sigma_{\partial X}(K) & \sigma_{\partial X}(Q) \end{pmatrix}(x', \xi') : \begin{array}{cc} H^s(\mathbb{R}_{x_n,+}) & H^{s-\mu}(\mathbb{R}_{x_n,+}) \\ \oplus & \to \quad \oplus \\ \pi_{\partial X}^*(J^-) & \pi_{\partial X}^*(J^+) \end{array} \qquad (4.74)$$

In the latter relation we took the more general notation $\pi_{\partial X}^*(J^-)$ instead of $\mathbb{C}^N$. This construction may be done first for (x', ξ') varying on the unit cosphere bundle induced by $T^*(\partial X)$, smoothly depending on (x', ξ'), and then it can be extended by the chosen homogeneity to $T^*(\partial X \setminus 0)$, which shows, that we obtain indeed a symbol in the operator-valued set-up. In this construction we also see some difference between the differential trace condition in (4.69) and the present one, which is more "non-stanard", namely, in integral form, and in any case non-local.

Summing up, we established for boundary problems like (4.59) a principal symbolic structure, namely,

$$\sigma(\mathcal{A}) := \left(\sigma_X^{(\mu)}(\mathcal{A}), \sigma_{\partial X}^{(\mu)}(\mathcal{A})\right), \qquad (4.75)$$

consisting of the principal inner and the boundary symbol, and the bijectivity of which just determines the ellipticity of $\mathcal{A}$. This in turn entails the Freholm property of (4.59) and the existence of a parametrix. There are some other natural properties, for instance, under compositions, similar to those from the case of pseudo-differential operators on compact closed manifolds. More details may be found in the above-mentioned monographs, in particular, in the last Chapter of [68], together with a discussion of boundary value problems with global projection conditions in the joint monograph of X. Liu and B.-W. Schulze [44], concerning the case, that the bundles in (4.74) may be not pull-backs of some bundles over ∂X. The material is altogether a kind of guideline for the case of edge and corner problems, where the underlying manifold has corresponding singularities.

There are to be added some details, that we first dropped, in particular, so-called Green operators appearing as extra summands G in the upper left corners (briefly mentioned before). Those also belong to aspects on 2×2-block-matrix valued operators, appearing in the analysis of boundary value problems. As we know from the standard pseudo-differential calculus, we are interested, in particular, in the composition of such operators. Those generate elements of the kind $G = K \circ T$, where T is a trace operator from the first factor, K a potential operator from the second one. Such an operator G is a special so-called Green operator. The linear structure of the block-matrix space of operators gives rise to Green operators like $G = \sum_{j=1}^{N} K_j \circ T_j$ with summands of the mentioned type. In other words, the upper left corners of block matrices should contain, al least, entries of the form $A + G$ with A being a pseudo-differential operator as before, and a Green operator G. such G also should contribute to the boundary symbolic structure, including aspects on ellipticity of the corresponding elements of the enlarged block-matrix algebra. Concerning more details we also refer to the last Chapter of the monograph of B.-W. Schulze [68], containing many details on this topic.

Remark 4.17 The general form of Green operators (modulo smoothing ones) takes the form of convergent sums

$$G = \sum_{j=1}^{\infty} K_j \circ T_j,$$

rather than the above-mentioned finite sums. with boundary symbols

$$\sigma_{\partial X}(G)(x', \xi') := \sum_{j=1}^{\infty} \sigma_{\partial X}(K_j)(x', \xi') \circ \sigma_{\partial X}(T_j)(x', \xi').$$

This comes from the fact, that operators G can be defined as pseudo-differential operators on ∂X with local operator-valued symbols $g(x', \xi')$ taking values in $\mathcal{S}(\overline{\mathbb{R}}_+) \hat{\otimes}_\pi \mathcal{S}(\overline{\mathbb{R}}_+)$ with $\hat{\otimes}_\pi$ indicating the projective tensor product between the involved spaces. This entails (according to a lemma of A. Pietsch [52]), a representation $g(x', \xi') = \sum_{j=1}^{\infty} k_j(x', \xi') \circ t_j(x', \xi')$ with potential and trace symbols $k_j(x', \xi')$ and $t_j(x', \xi')$, respectively. This gives rise to the above-mentioned representation of G with $K_j = \mathrm{op}_{x'}(k_j)$ and $T_j = \mathrm{op}_{x'}(t_j)$. We do not sketch other details here, since the main purpose of this consideration is to see formal similarities (and differences) between edge problems and boundary problems. Summing up, we complete this subsection with some properties of the class of boundary problems, analogously to elliptic pseudo-differential operators on closed compact manifolds, see L. Boutet de Monvel's work [9] and the monographs S. Rempel and B.-W. Schulze [61] or X. Liu and B.-W. Schulze [44]. The latter exposition develops the concept of boundary value or edge problems with so-called global projection conditions, in special cases such a phenomenon is known through the work of M.F. Atiyah an R. Bott [3] or of M.F. Atiyah, V. Patodi and I.M. Singer [6–8]. Note also, that the paper of J. Seiler [76, 77] treats boundary value problems with global projection conditions and with parameters.

Theorem 4.18 *The above-mentioned* 2×2 *block-matrices of operators* (4.59) *form an algebra* $\mathfrak{A}$, *more precisely, those of the form*

$$\mathcal{A} = \begin{pmatrix} A + G & T \\ K & Q \end{pmatrix}, \tag{4.76}$$

including Green operators G, with principal interior and boundary symbols $\sigma^{(\mu)}(\mathcal{A})$ and

$$\sigma_{\partial X}^{(\mu)}(\mathcal{A})(x', \xi') = \begin{pmatrix} \sigma_{\partial X}^{(\mu)}(A) + \sigma_{\partial X}(G) & \sigma_{\partial X}(T) \\ \sigma_{\partial X}(K) & \sigma_{\partial X}(Q) \end{pmatrix} (x', \xi'). \tag{4.77}$$

respectively. In particular, $\mathcal{A}_1$, $\mathcal{A}_2 \in \mathfrak{A}$ implies $\mathcal{A}_1 \circ \mathcal{A}_2 \in \mathfrak{A}$, provided that $\mathrm{im}(\mathcal{A}_2) = \mathrm{dom}(\mathcal{A}_1)$, and the involved orders are the same, and we have $\sigma^{(2\mu)}(\mathcal{A}_1 \circ \mathcal{A}_2) = \sigma^{(\mu)}(\mathcal{A}_1)\sigma^{(\mu)}(\mathcal{A}_2)$ and $\sigma_{\partial X}^{(2\mu)}(\mathcal{A}_1 \circ \mathcal{A}_2) = \sigma_{\partial X}^{(\mu)}(\mathcal{A}_1) \circ \sigma_{\partial X}^{(\mu)}(\mathcal{A}_2)$.

Clearly, we have more general composition results, when we admit the operators $\mathcal{A} \in \mathfrak{A}$ to have different orders and consider them as continuous maps

$$\mathcal{A} : \begin{array}{ccc} C^\infty(X, E_1) & & C^\infty(X, E_2) \\ \oplus & \to & \oplus \\ C^\infty(\partial X, J^-) & & C^\infty(\partial X, J^+) \end{array} . \tag{4.78}$$

The latter relation refers to smooth sections in complex vector bundles E_1 and E_2, respectively, over X, in the upper left corners of $\mathcal{A}$, (those generalize the above scalar operators), while $J^\pm$ are of the above meaning. Now the generalization of Theorem 4.18 refers to operators $\mathcal{A}_1$, $\mathcal{A}_2$ of orders μ and ν, respectively, (where the pairs of bundles for $\mathcal{A}_2$ concern in the image E, J) for certain E and J, and in the pre-image E_1, J^-). Then in Theorem 4.18 the second operator is assumed to be of order ν, and 2μ is to be replaced by $\mu + \nu$. Thus we can speak about continuity of operators between Sobolev spaces of distributional sections

$$\mathcal{A}_1 : \begin{array}{ccc} H^{s-\nu}(X, E) & & H^{s-\nu-\mu}(X, E_2) \\ \oplus & \to & \oplus \\ H^{s-\nu}(\partial X, J) & & H^{s-\nu-\mu}(\partial X, J^+) \end{array} \tag{4.79}$$

and

$$\mathcal{A}_2 : \begin{array}{ccc} H^s(X, E_1) & & H^{s-\nu}(X, E) \\ \oplus & \to & \oplus \\ H^s(\partial X, J^-) & & H^{s-\nu}(\partial X, J) \end{array} \tag{4.80}$$

for sufficiently large s.

Remark 4.19 The precise control of orders of

$$\begin{pmatrix} A + G & T \\ K & Q \end{pmatrix} : \begin{array}{c} H^s(X, E_1) \\ \oplus \\ H^s(\partial X, J^-) \end{array} \to \begin{array}{c} H^{s-\mu}(X, E_2) \\ \oplus \\ H^{s-\mu}(\partial X, J^+) \end{array} \tag{4.81}$$

of the operators T, K, Q in the above boundary problem (4.81) is often not very essential, since those can easily be modified by multiplying the corresponding block-matrices by order reducing factors from the left of the form $\begin{pmatrix} \mathrm{id} & 0 \\ 0 & Q_1 \end{pmatrix}$ and from the right by $\begin{pmatrix} \mathrm{id} & 0 \\ 0 & Q_2 \end{pmatrix}$ for elliptic elements Q_1 and Q_2 on ∂X of suitable orders, see the above Theorem 4.15. Here id indicates various identity operators in the occurring Sobolev spaces on X.

4.5   The Laplace-Beltrami Operator on a Wedge

In this Subsection we give a survey on ideas of the calculus of pseudo-differential operators on a manifold B with edge Y and motivate the symbolic structure of edge-degenerate operators. Later on, we return to *"technical"* elements of the approach. First remember, that the Laplace-Beltrami operator on $\mathbb{R}_+ \times X$ for any compact Riemannian manifold $X \in \mathfrak{M}_0$ is of Fuchs type and has the form

$$r^{-2} a_0(r \frac{\partial}{\partial r}) := r^{-2} \left((r \frac{\partial}{\partial r})^2 + (\dim(X) - 1) r \frac{\partial}{\partial r} + \Delta_g \right) \tag{4.82}$$

with Δ_g being the Laplacian on X associated with the metric g. The Laplacian on $\mathbb{R}_+ \times X \times \mathbb{R}^q$ then takes the form

$$r^{-2} a_0(r \frac{\partial}{\partial r}) + \sum_{j=1}^{q} (\frac{\partial}{\partial y_j})^2, \tag{4.83}$$

or, on the level of (r, z, η)-depending operator-valued symbols $r^{-2} a_0(z) - |\eta|^2$ for $\eta = (\eta_1, \dots, \eta_q)$. It is convenient in this connection, to choose the representation

$$r^{-2}\left(a_0(z) - |r\eta|^2 \right). \tag{4.84}$$

For purposes below, we assume, that $a_0(z)$ smoothly depends on additional variables $r \in \overline{\mathbb{R}}_+$, $y \in \Omega$ for a coordinate neighborhood $\Omega \subset Y$. This belongs to the motivation of studying symbols of the form

$$a(y, \eta) \in S^\mu(\Omega \times \mathbb{R}^q; E, \tilde{E}) \tag{4.85}$$

referring to a pseudo-differential action between

$$E := \mathcal{K}^{s,\beta}(\mathbb{R}_+ \times X), \ \tilde{E} := \mathcal{K}^{s-\mu,\beta-\mu}(\mathbb{R}_+ \times X). \tag{4.86}$$

By carrying out this pseudo-differential action we can pass to operator-valued symbols with values in $\mathcal{L}(E, \tilde{E})$. Another class of operator-valued symbols is associated with the couple of spaces (4.86). Therefore, we see a similarity with the situation of the spaces $H^s(\mathbb{R}^{n+1})$ and $H^{s-\mu}(\mathbb{R}^{n+1})$ gives us a symbol of the form

$$\tilde{p}(r, y, \rho, \eta) := r^{-\mu} p(r, y, r\rho, r\eta), \tag{4.87}$$

cf. the expression (4.84). Here we assume, that

$$p(r, y, \tilde{\rho}, \tilde{\eta}) \in S^\mu(\overline{\mathbb{R}}_+ \times \Omega \times \mathbb{R}^{1+q}). \tag{4.88}$$

Note that this type of expressions is also the topic of investigations of the papers [27, 28] with many details, to be presented here. Those belong to the so-called pseudo-differential edge calculus, and to some extent they can be subsumed under the tools of operator-valued symbols, considered before. More material in this context may be found, for instance, in the work of Ch. Dorschfeldt [21], N. Dines [20], T. Krainer [41], and in numerous other papers, mentioned in the references below, see also the monographs [67, 68]. In particular, we refer to the Definitions 4.11 and 4.12 (and to various notation in this connection) for the spaces (4.86). To be more precise, the operator-valued symbols concern

$$a(y, \eta) := \mathrm{op}_r(\tilde{p})(y, \eta), \tag{4.89}$$

with (4.87) being introduced before, where op_r just indicates the pseudo-differential action with respect to the variable $r \in \mathbb{R}_+$, which is integrated away, together with the covariable ρ. This implicitly contains a pseudo-differential action with respect to variables and covariables (x, ξ), and we obtain altogether

$$a(y, \eta) \in S^\mu(\Omega \times \mathbb{R}^q; E, \tilde{E}), \tag{4.90}$$

with $E, \tilde{E}$ being defined in (4.86). For convenience we assume for the moment that $p(r, y, \rho, r\eta)$ in (4.87) vanishes for r outside a compact set of $\mathbb{R}_+$ and for y outside a compact set of Ω. Later on the condition concerning r will be replaced by a weaker assumption, when we replace op_r via a Mellin quantization by op_z with $z \in \mathbb{C}$. On the other hand, the condition on the support of $p(r, y, \rho, r\eta)$ with respect to y is natural, since Ω only plays the role of a coordinate neighborhood of a global edge Y, which is (in simplest cases) a smooth compact C^∞-manifold, and symbols $p(r, y, \rho, r\eta)$ contain a factor $\varphi \in C^\infty(\Omega)$ connected with a partition of unity on Y, subordinate to an open covering of Y by coordinate neighborhoods Ω, corresponding to charts $\Omega \to \mathbb{R}^q$.

4.6 The Global Ellipticity of Edge Problems

In the following we describe the structure of the edge calculus, which will be more and more specified by the material below. By virtue of the complexity of the topic we hope, that such an approach is helpful to understanding basics of the approach. In particular, we skip technicalities; in particular, we refer to analogues from the calculus of boundary problems.

Roughly speaking, we study an algebra of 2×2-block matrices $\mathcal{A}$ of pseudo-differential operators on a manifold B with edge Y, and we focus on ellipticity, both of operators on B in upper left corners, and so-called edge conditions with respect to Y, formulated in terms of all entries of $\mathcal{A}$ in a tubular neighborhood of Y. Ellipticity will be expressed by symbolic levels, associated with the strata of the underlying configuration.

We already noted, that the structure of the elliptic edge calculus is similar to Boutet de Monvel's theory $\mathfrak{B}$ of pseudo-differential boundary value problems; for brevity we are using such a terminology here, though elliptic problems in this context are well-known. Those deserve careful attention, see also contributions by many authors, see also references, given in this exposition. Let us give the idea of our approach in terms of an operator algebra, and formulate its symbolic structure, in the context of ellipticity and parametrices.

Elliptic edge pseudo-differential operators A on a (compact and oriented) "pseudo"-manifold B with edge Y (as before, simply called, a manifold with edge), will be considered in connection with elliptic edge conditions, belonging to an algebra $\mathfrak{A}$ of 2×2-block matrices of operators.

$$
\mathcal{A} = \begin{pmatrix} A + G & K \\ T & Q \end{pmatrix} + \mathcal{G} : \begin{array}{ccc} \mathcal{W}^{s,\beta}(B) & & \mathcal{W}^{s-\mu,\beta-\mu}(B) \\ \oplus & \rightarrow & \oplus \\ H^s(Y, J^-) & & H^{s-\mu}(Y, J^+), \end{array}
\tag{4.91}
$$

where $\mathcal{W}^{s,\beta}(B)$ denotes weighted edge Sobolev spaces on the manifold B with edge Y, and $H^s(Y, J^-)$ is the Sobolev space on Y of distributional sections in a complex smooth vector bundle J^- over Y. More properties of the structure will be reflected by writing (4.91) in the form

$$
\mathcal{A} = \mathcal{A}_\psi + \mathcal{G}
\tag{4.92}
$$

for some negligible block-matrix valued summand $\mathcal{G}$, compact and smoothing, and with some extra specific features (e.g., to have smoothing, holomorphic Mellin symbols). Concerning the interpretation of $J^\pm$, see also formula (4.93) below. Operators like (4.91) do constitute the edge algebra on B, cf. material of the following subsections. First we will sketch some motivation of this relatively voluminous program.

Once we refer to the notion of a manifold with edge (which is an analogue of a manifold with boundary, here in the context of "pseudo-manifolds"), we establish the concept of ellipticity of elements in the above-mentioned algebra, which will be of similar meaning as standard ellipticity of pseudo-differential operators on a smooth manifold, based on symbolic structuctures, with the existence of a parametrix within the algebra, the Fredholm property in adequate analogues of Sobolev spaces and elliptic regularity.

This program will be realised in a similar manner as for boundary value problems on smooth manifolds with boundary, see the work of L. Boutet de Monvel [9], where ellipticity of the upper left corner A_{11} on $B \setminus Y$ is considered together with the ellipticity of $\mathcal{A}$ near the edge Y (similarly to the ellipticity of boundary conditions, to be required in boundary value problems). The ellipticity of A takes place in terms of a non-vanishing principal symbol $\sigma_\psi^{(\mu)}(A)$ on $T^*B \setminus 0$, called interior symbol of $\mathcal{A}$. In a tubular neighborhood V of Y in B we associate with $\sigma_\psi^{(\mu)}(A)$ a so-called edge symbol $\sigma_\wedge^{(\mu)}(A)$, which is a (y, η)-depending family of Fredholm operators,

$$\sigma_\wedge^{(\mu)}(A)(y, \eta) : E \to \tilde{E}$$

for $E := \mathcal{K}^{s,\beta}(\mathbb{R}_+ \times X)$, $\tilde{E} := \mathcal{K}^{s-\mu,\beta-\mu}(\mathbb{R}_+ \times X)$, cf. (4.11). Now we generate a block-matrix valued edge symbol

$$\sigma_\wedge^{(\mu)}(\mathcal{A})(y, \eta) = \begin{pmatrix} \sigma_\wedge^{(\mu)}(A + G) & \sigma_\wedge(T) \\ \sigma_\wedge(K) & \sigma_\wedge(Q) \end{pmatrix} (y, \eta) : \begin{matrix} E \\ \oplus \\ \pi_Y^*(J^-) \end{matrix} \to \begin{matrix} \tilde{E} \\ \oplus \\ \pi_Y^*(J^+) \end{matrix} , \tag{4.93}$$

which is an isomorphism for some finite-dimensional (complex) vector bundles J^-, J^+ on Y. They play a similar role as those in the book of X. Liu and B.-W. Schulze [45], see also S. Rempel [59]. Moreover, π_Y^* is the bundle pull-back under the canonical projection $\pi_Y : T^*(Y) \setminus 0 \to Y$ with $T^*(Y)$ being the cotangent bundle of Y, and 0 indicating the zero-section (identified with Y). The upper index $^{(\mu)}$ in the above formula indicates adequate homogeneities of the ingredients.

From the pair of symbols

$$\left(\sigma_\psi^\mu(A), \sigma_\wedge^{(\mu)}(\mathcal{A}) \right) \tag{4.94}$$

(also written as $\left(\sigma_\psi^\mu(A)(x, \xi), \sigma_\wedge^{(\mu)}(\mathcal{A})(y, \eta) \right)$) we obtain an associated block-matrix of operators of the form (4.91), which is called elliptic, when both components of (4.94) are elliptic. The ellipticity of $\mathcal{A}$ entails the existence of a parametrix $\mathcal{P}$ of $\mathcal{A}$ and also the Fredholm property of (4.91) for every s. As mentioned before, the papers [27, 28] contain many further details of the edge calculus. This mainly concerns the calculus close to the edge Y of B in terms of operator-valued symbols, and other specific aspects. Another part of the of the calculus on B consists of the "usual" pseudo-differential analysis on $B \setminus Y$.

This will be tacitly used. Then the operators close to Y are to be glued together with those on $B \setminus Y$ by using a partition of unity, subordinate to an open covering of B by open sets Ω_j for $j = 1, 2$, where Ω_1 is an open subset of $B \setminus Y$ and Ω_2 a tubular neighborhood of Y in B. The details are more or less straightforward, provided that we can be sure, that the different contributions to the calculus from Ω_1 and Ω_2 are compatible over the intersection of both open sets. However, this is the case, as we have seen before. Therefore, we may concentrate on the contributions from Ω_2.

4.7   Local Calculus Close to the Edge

This subsection will outline some elements of the calculus close to the edge. Details are motivated by considerations of the preceding subsections. Starting point are operator functions

$$r^{-\mu} p(r, y, r\rho, r\eta) \tag{4.95}$$

defined in terms of

$$p(r, y, \tilde{\rho}, \tilde{\eta}) \tag{4.96}$$

satisfying (4.88) and taking values in $\mathcal{L}(E, \tilde{E})$ for some Hilbert spaces $E, \tilde{E}$ endowed with group actions $\{\kappa_\lambda\}_{\lambda \in \mathbb{R}_+}$ and $\{\tilde{\kappa}_\lambda\}_{\lambda \in \mathbb{R}_+}$, respectively. Symbols of this kind have been studied in detail in [27]. Here we are mainly interested in an operator convention, where the dependence on $r\rho$ turns to a dependence on the complex variable $z \in \mathbb{C}$.

Definition 4.20 By

$$S^\mu(\overline{\mathbb{R}}_+ \times \Omega \times \mathbb{C} \times \mathbb{R}^q; E, \tilde{E}) \tag{4.97}$$

we denote the set of all

$$h(r, y, z, \eta) \in C^\infty(\overline{\mathbb{R}}_+ \times \Omega \times \mathbb{C} \times \mathbb{R}^q, \mathcal{L}(E, \tilde{E}))$$

that are holomorphic in $z \in \mathbb{C}$ with values in $S^\mu(\overline{\mathbb{R}}_+ \times \Omega \times \mathbb{R}^q; E, \tilde{E})$ and

$$h(r, y, z, \eta) \in S^\mu(\overline{\mathbb{R}}_+ \times \Omega \times \Gamma_\alpha \times \mathbb{R}^q; E, \tilde{E})$$

for every real α, uniformly in every compact α-interval.

In an analogous manner we define classical symbols, indicated by subscript "cl" at (4.97).

Theorem 4.21 *For every*

$$p(r, y, r\rho, \eta) \in S^{\mu}(\overline{\mathbb{R}}_+ \times \Omega \times \mathbb{R}^{1+q}; E, \tilde{E})$$

there exists an

$$h(r, y, z, \eta) \in S^{\mu}(\overline{\mathbb{R}}_+ \times \Omega \times \mathbb{C} \times \mathbb{R}^q; E, \tilde{E})$$

such that

$$\mathrm{op}_r(p)(y, \eta) = \mathrm{op}_M^{\frac{1}{2}}(h)(y, \eta) \tag{4.98}$$

mod $C^{\infty}(\Omega, L^{-\infty}(\mathbb{R}_+ \times \mathbb{R}^q; E, \tilde{E}))$, *and h has the asymptotic expansion*

$$h(r, y, i\rho, \eta) \sim p(r, y, -r\rho, \eta) + \sum_{k=1}^{\infty} \sum_{j=0}^{k} c_{k,j} \partial_\rho^{k+j} p(r, y, -r\rho, \eta) \rho^j \tag{4.99}$$

for p-independent real constants $c_{k,j}$, where the asymptotic summation is carried out in $S^{\mu}(\overline{\mathbb{R}}_+ \times \Omega \times \mathbb{R}^{1+q}; E, \tilde{E})$. An analogous result holds for classical symbols with corresponding asymptotic expansions for all homogeneous components.

The proof of Theorem 4.21 may be found in [27]. It shows, in particular, that a suitable quantization of the symbol p, containing the covariable $r\rho$ yields a holomorphic Mellin symbol with the respective covariable z.

Remark 4.22 Note that the procedure of kernel cut-off gives us many candidates of holomorphic Mellin symbols, representing $\mathrm{op}_r(p)$ modulo some smoothing remainder, since kernel cut-off contains an arbitrary cut-off function φ.

In order to see more details on the kernel cut-off, we briefly consider certain operator-valued holomorphic Mellin symbols and outline some useful information, cf. [28]. Recall that

$$M_O^{\mu}(X; \mathbb{R}^q) := \mathcal{A}(\mathbb{C}, L_{\mathrm{cl}}^{\mu}(X; \mathbb{R}^q)). \tag{4.100}$$

For $q = 0$ we simply write $M_O^{\mu}(X)$.

Definition 4.23 Define

$$C_{\mathrm{deg}}^{\infty}(\overline{\mathbb{R}}_+ \times \Omega, L_{\mathrm{cl}}^{\mu}(X; \mathbb{R}^{1+q})) := \{p := p(r, y, \rho, \eta) = \tilde{p}(r, y, r\rho, r\eta)\}$$

for

$$p \in C^\infty(\overline{\mathbb{R}}_+ \times \Omega, L^\mu_{\mathrm{cl}}(X; \mathbb{R}^{1+q}))$$

and

$$C^\infty_{\mathrm{deg}}(\overline{\mathbb{R}}_+ \times \Omega, M^\mu_O(X; \mathbb{R}^q)) := \{h := h(r, y, z, \eta) = \tilde{h}(r, y, z, r\eta)\}$$

for

$$h \in C^\infty(\overline{\mathbb{R}}_+ \times \Omega, M^\mu_O(X; \mathbb{R}^q)).$$

Theorem 4.24 *Assume that*

$$p \in C^\infty(\overline{\mathbb{R}}_+ \times \Omega, L^\mu_{\mathrm{cl}}(X; \mathbb{R}^{1+q})).$$

Let $\varphi \in C_0^\infty(\mathbb{R}_+)$ satisfy the condition $\varphi \equiv 1$ in a neighborhood of 1. Then there exists an $h \in C^\infty(\overline{\mathbb{R}}_+ \times \Omega, M^\mu_O(X; \mathbb{R}^q))$ such that

$$\mathrm{op}_r(p)(y, \eta) - \mathrm{op}_M^{\frac{1}{2}}(h)(y, \eta) = \mathrm{op}_r(Q)(y, \eta) \in C^\infty(\Omega_y, L^{-\infty}(\mathbb{R}_+ \times X; \mathbb{R}^q_\eta))$$

for $Q(r, r', y, \eta) = (1 - \varphi(r'/r))p(r, y, \rho, \eta)$, $r, r' \in \mathbb{R}_+$, $(\rho, \eta) \in \mathbb{R}^{1+q}$.

A proof of this result may be found in [28], Section A1.

Given p and h as in Theorem 4.24 we call h a Mellin quantization of p. Let us set

$$p_\wedge := \tilde{p}(0, y, r\rho, r\eta), \ h_\wedge := \tilde{h}(0, y, z, r\eta). \tag{4.101}$$

Recall that we have the spaces $\mathcal{H}^{s,\beta}(X^\wedge)$, defined in terms of $\mathcal{H}^{s,\beta}(\mathbb{R}_+ \times \mathbb{R}^n)$, where the latter space is the closure of $C_0^\infty(\mathbb{R}_+ \times \mathbb{R}^n)$ with respect to the norm

$$\|u\|^2_{\mathcal{H}^{s,\beta}(\mathbb{R}_+ \times \mathbb{R}^n)} = \int_{\Gamma_{\frac{1+n}{2} - \beta}} \int_{\mathbb{R}^n} (1 + |z|^2 + |\xi|^2)^s (|(M_{r \to z} F_{x \to \xi} u)(x, \xi)|^2) |dz| d\xi.$$

$$\tag{4.102}$$

The space $\mathcal{H}^{s,\beta}(X^\wedge)$ is then defined as the completion of $C_0^\infty(X^\wedge)$ with respect to the norm

$$\|u\|^2_{\mathcal{H}^{s,\beta}(X^\wedge)} = \sum_{j=1}^N \left\| (\varphi_j u) \circ (1 \times \theta_j)^{-1} \right\|^2_{\mathcal{H}^{s,\beta}(\mathbb{R}_+ \times \mathbb{R}^n)} \tag{4.103}$$

with $(\varphi_j)_{j=1,\dots,N}$ being a partition of unity subordinate to an open covering of X by coordinate neighborhoods $(U_j)_{j=1,\dots,N}$ and charts $\theta_j : U_j \to V_j \subset \mathbb{R}^n$.

Moreover, recall the definition of spaces $\mathcal{K}^{s,\beta}(X^\wedge)^\rho$ by the norm

$$\|u\|^2_{\mathcal{K}^{s,\beta}(X^\wedge)^\rho} = \|\omega u\|^2_{\mathcal{H}^{s,\beta}(X^\wedge)} + \sum_{j=1}^{N} \left\| [(1-\omega)\varphi_j u] \circ \kappa_j^{-1} \right\|^2_{\mathcal{H}^{s,\rho}(\mathbb{R}^{1+n})} \tag{4.104}$$

for $\kappa_j(r, x) := r\theta_j(x) : \mathbb{R}_+ \times U_j \to \mathbb{R}^{1+n}$ with the above-mentioned φ_j and θ_j.

Note, in particular, that $\mathcal{H}^{0,0}(X^\wedge) = \mathcal{K}^{0,0}(X^\wedge) = r^{\frac{n+1}{2}} L^2(X^\wedge)$. Set $\mathcal{K}^{s,\beta}(X^\wedge)^\rho :=$ $r^\rho \mathcal{K}^{s,\beta}(X^\wedge)$. Then, we have $r^{-\mu} \mathcal{K}^{s,\beta}(X^\wedge)^\rho = \mathcal{K}^{s,\beta-\mu}(X^\wedge)^{\rho+\mu}$.

Definition 4.25 A symbol $g(y, \eta)$ is called Green, associated with the pair of weights β, β', if

$$g(y, \eta) \in \bigcap_{s,s',\rho,\rho' \in \mathbb{R}} S^\mu_{\mathrm{cl}}(\Omega \times \mathbb{R}^q; \mathcal{K}^{s,\beta}(X^\wedge)^\rho, \mathcal{K}^{s',\beta'}(X^\wedge)^{\rho'})$$

and

$$g^*(y, \eta) \in \bigcap_{s,s',\rho,\rho' \in \mathbb{R}} S^\mu_{\mathrm{cl}}(\Omega \times \mathbb{R}^q; \mathcal{K}^{s,-\beta'}(X^\wedge)^\rho, \mathcal{K}^{s',-\beta}(X^\wedge)^{\rho'}).$$

4.8   Global Ellipticity with Edge Conditions

The calculus on a manifold B with edge Y is an analogue of boundary-value problems, where B is a manifold with boundary Y. Elliptic operators are then connected with boundary conditions. Those may be of trace- or potential type, cf. [9] or [24], see the block-matrices (4.93). Here we develop some details of the corresponding calculus. The method will be, that we specify the details around constructing elliptic block-matrices of the shape (4.93) in the framework of the edge calculus. This can be done by different methods. On the compact manifold B with smooth edge Y of dimension q we first fix an open neighborhood Ω on Y and a tubular neighborhood of Y in B such that the symbol $r^{-\mu}a_{\mathrm{int}}(r, x, y, r\rho, \xi, r\eta)$ of order μ of the edge-degenerate operator A in consideration is elliptic on V. For technical reasons we assume, that V has the shape of a rectangle with respect to local variables $0 < r < c$ for some $c > 0$ and $y \in \Omega$. We then pass to local coordinates, such that the interval $0 < r < c$ is identified with $\mathbb{R}_+ \ni r$. Later on, globally on B the operators are constructed via a partition of unity with suitable cut-off functions, subordinate to an open covering of $B \setminus Y$, consisting of the collar neighborhood of Y, together with a remaining open subset of $B \setminus V$. We mainly focus on effects on the collar neighborhood of Y, while operators over $B \setminus V$ are elliptic in the usual sense. We first sketch the method of constructing operators of the calculus along ideas from boundary value problems of the preceding section. After that, we consider specific novelties in connection with edge case. The above-mentioned local symbol a is assumed to be elliptic for $r > 0$. Applying a kernel cut-off procedure which only leaves compact Green remainders, we may assume, that the symbol depends on the covariable z rather than $r\rho$. In addition we pass to a corresponding (holomorphic in z) operator-valued symbol and actions on $\mathcal{K}^{s,\beta}(\mathbb{R}_+ \times X)$-spaces, where $X \in \mathfrak{M}_0$ is compact, and we write $a_\beta(r, y, z, \eta)$. We start with the assumption, that

$$\mathrm{op}_r(a_\beta)(y, \eta) : \mathcal{K}^{s,\beta}(\mathbb{R}_+ \times X) \to \mathcal{K}^{s-\mu,\beta-\mu}(\mathbb{R}_+ \times X) \tag{4.105}$$

is a family of elliptic operators of Fuchs type close to $r = 0$ for $y \in \Omega$, $\eta \in \mathbb{R}^q$, which is in addition exit-elliptic up to $r \to \infty$.

Globally on B and modulo lower order terms the shape of the operators (4.93) corresponds to operator block-matrices

$$\mathcal{A} := \begin{pmatrix} A + G & K \\ T & Q \end{pmatrix} : \begin{matrix} \mathcal{W}^{s,\beta}(B) \\ \oplus \\ H^s(Y, J^-) \end{matrix} \to \begin{matrix} \mathcal{W}^{s-\mu,\beta-\mu}(B) \\ \oplus \\ H^{s-\mu}(Y, J^+) \end{matrix} , \tag{4.106}$$

see (4.91). Here

$$\mathcal{W}^{s,\beta}(B) := \omega_B \mathcal{W}^s(Y, \mathcal{K}^{s,\beta}(\mathbb{R}_+ \times X)) + (1 - \omega_B) H^s_{\mathrm{loc}}(B \setminus Y)$$

are weighted edge Sobolev spaces on the manifold B with edge Y. Here $\omega_B \in C^\infty(\overline{\mathbb{R}}_+)$ is a cut-off function on B which is supported in V with $\omega_B \equiv 1$ near Y. The operator A is a classical pseudo-differential operator in $B \setminus Y$ of order μ, and by $\sigma_\psi^{(\mu)}(A)$ we denote its homogeneous principal symbol. The operator A is locally close to Y of the form $\omega_B \mathrm{op}_y \mathrm{op}_r(a_\beta)\tilde{\omega}_B$ for cut-off functions $\omega_B, \tilde{\omega}_B$ of the above-mentioned kind, such that $\omega_B \equiv 1$ on $\mathrm{supp}(\tilde{\omega}_B)$ where $\mathrm{op}_r(a_\beta)$ is given by (4.105), and G is a Green operator. Furthermore, $H^{s,\beta}(Y, J)$ are Sobolev spaces of distributional sections in vector bundles J over Y of smoothness s.

$$T : \mathcal{W}^{s,\beta}(B) \to H^{s-\mu}(Y, J^+)$$

is a trace operator with a homogeneous principal boundary symbol $\sigma_\partial^{(\mu)}(T)(y, \eta) :$ $\mathcal{K}^{s,\beta}(\mathbb{R}_+ \times X) \to \pi_Y^*(J^+)$,

$$K : H^s(Y, J^-) \to \mathcal{W}^{s-\mu,\beta-\mu}(B)$$

is a potential operator with a homogeneous principal boundary symbol $\sigma_\partial^{(\mu)}(K)(y, \eta)$:
$\pi_Y^*(J^-) \to \mathcal{K}^{s-\mu, \beta-\mu}(\mathbb{R}_+ \times X)$, and

$$Q : H^s(Y, J^-) \to H^{s-\mu}(Y, J^+)$$

is a classical pseudo-differential operator on Y of order μ with the homogeneous principal
symbol $\sigma_\partial^{(\mu)}(Q)(y, \eta) : \pi_Y^*(J^-) \to \pi_Y^*(J^+)$. The operator G can be written in form of a
convergent series $G = \sum_{j=0}^\infty K_j \circ T_j$ for trace operators T_j and potential operators K_j with
the homogeneous principal boundary symbol $\sigma_\partial^{(\mu)}(G)(y, \eta) = \sum_{j=0}^\infty \sigma_\partial^{(\mu)}(K_j) \circ \sigma_\partial^{(\mu)}(T_j)$.
The scheme of considerations corresponds to that of L. Boutet de Monvel [9] and G.I.
Eskin [24], for the special case of boundary value problems, see also the monographs of
G. Grubb [30], S. Rempel and B.-W. Schulze [59] and the work of E. Schrohe [64] .

Let us now turn to ellipticity of operators (4.106). First we assume, that A is elliptic on
B, i.e., A is elliptic on $B \setminus Y$ as an elliptic pseudo-differential operator in the usual sense.
Remember, that the above-mentioned operator $\omega_B \mathrm{op}_y \mathrm{op}_r(a_\beta) \tilde{\omega}_B$ close to Y is elliptic on V
and that $\mathrm{op}_r(a_\beta)(y, \eta)$ for every fixed (y, η) is Fredholm between the respective $\mathcal{K}$-spaces
on $\mathbb{R}_+ \times X$. For sufficiently large N we find an operator-valued symbol

$$k(y, \eta) \in S^\infty(\Omega \times \mathbb{R}^q; \mathbb{C}^N, \mathcal{K}^{s-\mu, \beta-\mu}(\mathbb{R}_+ \times X))$$

such that

$$\left(\mathrm{op}_r(a_\beta) \; k\right)(y, \eta) : \left(\mathcal{K}^{s,\beta}(\mathbb{R}_+ \times X) \; \mathbb{C}^N\right) \to \mathcal{K}^{s-\mu, \beta-\mu}(\mathbb{R}_+ \times X) \tag{4.107}$$

is a surjective map for all $(y, \eta) \in \Omega \times \mathbb{R}^q$, where the second component $k(y, \eta)$ represents
a symbol in (y, η) of suitable order, referring to the group action in $\mathcal{K}^{s-\mu, \beta-\mu}(\mathbb{R}_+ \times X)$.
The trivial bundle $\Omega \times \mathbb{C}^N$ will be interpreted as J^-, and locally we define

$$K := \mathrm{op}_y k : H^s_{\mathrm{comp}}(\Omega, J^-) \to \mathcal{W}^{s-\mu}_{\mathrm{loc}}(\Omega, \mathcal{K}^{s-\mu, \beta-\mu}(\mathbb{R}_+ \times X)).$$

The operators $k(y, \eta)$ are of finite rank, and the operators (4.107) are surjective for all
(y, η). Thus the family of kernels of (4.107) form a finite-dimensional vector bundle over
$T^*(Y)$, the fibres $b(y, \eta)$ of which are smooth. We assume in the sequel, that this bundle
is the pull-back of some bundle J^+ on Y. Thus we can interpret $b(y, \eta)$ as a pseudo-
differential symbol, inducing a continuous operator

$$\left(T \; Q\right) : \ker\left(\mathrm{op}_r(a_\beta) \; k\right) \to H^{s-\mu}(Y, J^+). \tag{4.108}$$

We may interpret (4.108) also as the second row of a continuous operator (4.106) for some
pseudo-differential operator $Q : H^s(Y, J^-) \to H^{s-\mu}(Y, J^+)$.

Theorem 4.26 *Let*

$$
\mathcal{A}_1 := \begin{pmatrix} A_1 + G_1 & K_1 \\ T_1 & Q_1 \end{pmatrix} : \begin{array}{c} \mathcal{W}^{s-\mu,\beta-\mu}(B) \\ \oplus \\ H^{s-\mu}(Y, J) \end{array} \to \begin{array}{c} \mathcal{W}^{s-(\mu+\nu),\beta-(\mu+\nu)}(B) \\ \oplus \\ H^{s-(\mu+\nu)}(Y, J^+) \end{array} , \qquad (4.109)
$$

and

$$
\mathcal{A}_2 := \begin{pmatrix} A_2 + G_2 & K_2 \\ T_2 & Q_2 \end{pmatrix} : \begin{array}{c} \mathcal{W}^{s,\beta}(B) \\ \oplus \\ H^{s}(Y, J^-) \end{array} \to \begin{array}{c} \mathcal{W}^{s-\mu,\beta-\mu}(B) \\ \oplus \\ H^{s-\mu}(Y, J) \end{array} , \qquad (4.110)
$$

be operators analogous to (4.106) on a manifold B with edge Y. Then the composition

$$
\mathcal{A}_1\mathcal{A}_2 : \begin{array}{c} \mathcal{W}^{s,\beta}(B) \\ \oplus \\ H^{s}(Y, J^-) \end{array} \to \begin{array}{c} \mathcal{W}^{s-(\mu+\nu),\beta-(\mu+\nu)}(B) \\ \oplus \\ H^{s-(\mu+\nu)}(Y, J^+) \end{array} \qquad (4.111)
$$

is again of analogous structure, and we have the symbolic rules

$$
\sigma_\psi^{(\nu+\mu)}(A_1 A_2) = \sigma_\psi^{(\nu)}(A_1)\sigma_\psi^{(\mu)}(A_2), \ \sigma_\wedge^{(\nu+\mu)}(\mathcal{A}_1\mathcal{A}_2) = \sigma_\wedge^{(\nu)}(\mathcal{A}_1)\sigma_\wedge^{(\mu)}(\mathcal{A}_2). \qquad (4.112)
$$

The proof is an immediate consequence of the definitions. Theorem 4.26 allows us to speak about the edge algebra $\mathfrak{B}$ of corresponding block matrix operators over the manifold B with edge Y. Let us now turn to the concept of ellipticity and of parametrices within $\mathfrak{B}$.

Definition 4.27 An operator $\mathcal{A} \in \mathfrak{B}$ is said to be elliptic, if $\sigma_\psi^{(\mu)}(A)(x, \xi)$ and $\sigma_\wedge^{(\mu)}(\mathcal{A})(y, \eta)$. are bijective for all $x \in B \setminus Y$ and $\xi \neq 0$ and for all $y \in Y$ and $\eta \neq 0$, respectively.

Recall. that $\sigma_\psi^{(\mu)}(A)(x, \xi)$ is the standard homogeneous principal symbol of order μ of the classical pseudo-differential operator A on the corresponding smooth manifold $\Omega = B \setminus Y$, while

$$
\sigma_\wedge^{(\mu)}(\mathcal{A})(y, \eta) = \begin{pmatrix} \sigma_\wedge^{(\mu)}(A + G) & \sigma_\wedge(T) \\ \sigma_\wedge(K) & \sigma_\wedge(Q) \end{pmatrix} (y, \eta) : \begin{array}{c} E \\ \oplus \\ \pi_Y^*(J^-) \end{array} \to \begin{array}{c} \tilde{E} \\ \oplus \\ \pi_Y^*(J^+) \end{array} , \qquad (4.113)
$$

is the operator-valued homogeneous principal symbol of the block-matrix operator $\mathcal{A}$ in a collar neighborhood of Y in $B \setminus Y$, for

$$
\sigma_\wedge^{(\mu)}(\mathcal{A})(y, \eta) : E \to \tilde{E} \qquad (4.114)
$$

for $E := \mathcal{K}^{s,\beta}(\mathbb{R}_+ \times X)$, $\tilde{E} := \mathcal{K}^{s-\mu,\beta-\mu}(\mathbb{R}_+ \times X)$, cf. (4.11). Other ingredients of the above block-matrix valued edge symbol $\sigma_\wedge^{(\mu)}(\mathcal{A})(y,\eta)$ refer to finite-dimensional vector bundles J^-, J^+ on Y. Note, that the operators $\sigma_\wedge^{(\mu)}(G)(y,\eta)$, $\sigma_\wedge(K)(y,\eta), \sigma_\wedge(T)(y,\eta), \sigma_\wedge(Q)(y,\eta)$ just represent extra elliptic edge-conditions, analogously to elliptic boundary conditions to an elliptic operator A in Boutet de Monvel's calculus of pseudo-differential boundary value problems, mentioned before.

The following theorem for elliptic operators in the edge-calculus (according to the above Definition 4.27), is an analogue of a corresponding result on elliptic operators in Boutet de Monvel's algebra.

Theorem 4.28 *Let $\mathcal{A} \in \mathfrak{B}$ be an elliptic edge problem in the sense of Definition 4.27 . Then there exists an operator $\mathcal{P} \in \mathfrak{B}$ which is again elliptic in that sense, which is a parametrix of $\mathcal{A}$, such that*

$$\mathcal{A}\mathcal{P} = \mathcal{I}_1 - \mathcal{K}_1, \tag{4.115}$$

and

$$\mathcal{P}\mathcal{A} = \mathcal{I}_2 - \mathcal{K}_2 \tag{4.116}$$

for compact operators $\mathcal{K}_1, \mathcal{K}_2 \in \mathfrak{B}$.

Proof The proof may be obtained by first choosing an elliptic operator $\mathcal{P}_0 \in \mathfrak{B}$ with $\sigma_\psi^{(-\mu)}(\mathcal{P}_0)\sigma_\psi^{(\mu)}(A)^{-1}$ with arbitrary elliptic conditions. Then the operator $\mathcal{P}_1 := \mathcal{A}\mathcal{P}_0 \in \mathfrak{B}$ is elliptic in $\mathfrak{B}$ and satisfies $\sigma_\psi^{(0)}(\mathcal{P}_1) = 1$. In a next step we construct an elliptic operator $\mathcal{P}_2 \in \mathfrak{B}$ which is a parametrix of $\mathcal{P}_1$. The ideas for computing $\mathcal{P}_2$ may be found in the monograph of X. Liu and B.-W. Schulze [44]. In a final step, using Theorem 4.26, we may set $\mathcal{P} := \mathcal{P}_0\mathcal{P}_2$. $\qquad\square$

The Geometry of Singularities

5

5.1   The Case of Conical Singularities or Edges

Details on distribution spaces and pseudo-differential operators on configurations with singularities are to a large extent determined by the underlying geometry, see the material in the preceding sections. We may observe similar interactions between analysis an geometry in the context of higher edges or corners. The insight from elementary cone and edge operators is a source of ideas for the higher corner and edge case. In the following we focus on pseudo-manifolds, mainly called manifolds with singularities. Those are also of interest in combination with exits to infinity in the sense of, for example, H.O. Cordes [18]. In general, by a manifold with singularities we understand a set M, containing a singular subset M_{sing}, such that $M_{\mathrm{int}} := M \setminus M_{\mathrm{sing}}$ is a smooth Riemannian manifold, and M is altogether defined by repeatedly forming cones and wedges (to be described in more detail below). For later constructions we choose an open neighborhood U of M_{sing} in M; clearly $U \setminus M_{\mathrm{sing}}$ is also smooth.

Analogously to the above-mentioned way of forming manifolds with conical singularities we can proceed for higher singularities. Let us start with the case of a compact manifold B with a smooth edge Y. In order to interpret B as the link of a certain infinite cone, we first observe, that B can be regarded as a subset of the unit sphere $\mathbb{S}^n$ in the Euclidean space $\mathbb{R}^{n+1}$ for sufficiently large n. In other words, we interpret the manifold with edge B as a subset of $\mathbb{S}^n$ and use for convenience the same notation. In a next step we form B^Δ, defined as the union of all infinite straight half-lines in $\mathbb{R}^{n+1}$, starting at the origin $0 \in \mathbb{R}^{n+1}$ and going through any $b \in B$. Applying a diffeomorphism $\Phi : \mathbb{R}^{n+1} \to \mathbb{R}^{n+1}$ and setting $C := \Phi(B^\Delta)$, we obtain another manifold with corner singularity $c := \Phi(0)$. In this way we obtain a more general example of a manifold with corner singularity c and link $\Phi(B)$. Note, that we have the subcone $Y^\Delta \subset B^\Delta$. Then $M_{\mathrm{reg}} := B^\Delta \setminus Y^\Delta$ is a smooth Riemannian manifold contained in $M := B^\Delta$ with singular

© Springer Nature Switzerland AG 2025
D.-C. Chang, B.-W. Schulze, *Analysis on Manifolds with Singularities*,
Pseudo-Differential Operators 16, https://doi.org/10.1007/978-3-032-01890-8_5

subset $M_{\text{sing}} := Y^\triangle$. By construction $\Phi(B^\triangle \setminus Y^\triangle)$ is a smooth Riemannian manifold with a degenerate behavior of the metric at the singular subset $\Phi(M_{\text{sing}}) = \Phi(Y^\triangle)$. In particular, we described the geometry of the singularity of an infinite straight cone with link B. In other words, we have defined the geometry of a manifold with (second order) corner. It is now evident, how to organize an infinite straight cone with link $N \in \mathfrak{M}_k$, where $k = 1$ corresponds to a link with conical singularity or edge.

5.2   Higher Singularities

The case of $\mathfrak{M}_k$ for $k > 1$ is iteratively defined as mentioned above, and we distinguish between manifolds with corners or edges of order k. There are lots of interesting observations on the categories $\mathfrak{M}_k$, but here we content ourselves with examples and certain elementary properties. Consider a compact element $M \in \mathfrak{M}_1$ which is a configuration with edge Y. Then it is common to pass to the stretched manifold $\mathbb{M}$ of M by replacing a tubular neighborhood U of Y by a stretched neighborhood $\mathbb{U}$, see the explanation below, and then to paste $\mathbb{U}$ to $M \setminus U$ in an evident manner, to obtain $\mathbb{M}$. In order to explain $\mathbb{U}$ in terms of U we identify U with the set $Y \times X_1^\triangle$, where by $X_1^\triangle$ we understand the piece of $X^\triangle$ defined by $\{(x, t) : x \in X, 0 < t \leq 1\} \cup c$, with c indicating the tip of $X^\triangle$. Note, that there is a canonical map $X \times [0, 1] \to X_1^\triangle$. Then we simply set $\mathbb{U} := X \times [0, 1]$. Summing up, we completed the construction of the stretched manifold $\mathbb{M}$ in the case $M \in \mathfrak{M}_1$ and $X \in \mathfrak{M}_0$.

It is now clear, how the above-mentioned iteration works for higher singularity orders. For the iterative construction we assume $X \in \mathfrak{M}_1$ and define elements $M \in \mathfrak{M}_2$ by looking at corners $X^\triangle$ or $Y \times X^\triangle$ for any second order edge $Y \in \mathfrak{M}_0$. Then $M \in \mathfrak{M}_2$ is a pseudo-manifold with an exceptional set $M_{\text{sing}} := Y$ (Y may be of dimension 0). The above process, applied to $M := \mathbb{X}$ gives us the associated stretched manifold $\mathbb{M}$. Then the iteration can be continued. The operators living on $M \in \mathfrak{M}_k$ will refer to weighted distributions on the corresponding stretched manifolds, and those are defined in terms of the partition of unity on $\mathbb{M}$ subordinated to the open covering of $\mathbb{M}$ for $M \in \mathfrak{M}_k$ by $\mathbb{U}_{\text{sing}}$ and U, where the latter open set has been defined before, and we form

$$\mathbb{U}_{\text{sing}} \cup \tilde{M}_{\text{reg}} = \mathbb{M}. \tag{5.1}$$

where $\tilde{M}_{\text{reg}}$ is a submanifold of M_{reg} such that $\tilde{M}_{\text{reg}} \cup M_{\text{sing}} = M$. Let us return for a moment to $M \in \mathfrak{M}_k$. Then we have $M_{k-1} := M_{\text{sing}} \in \mathfrak{M}_{k-1}$, moreover, $M_{k-2} := (M_{k-1})_{\text{sing}} \in \mathfrak{M}_{k-2}$, and so on. This gives us a decreasing chain of subsets $M := M_k, M_{k-1}, M_{k-2}, \ldots, M_0$ with $M_j \in \mathfrak{M}_j$, in particular, $M_0 \in \mathfrak{M}_0$. Given elements $M, N \in \mathfrak{M}_k$, then a morphism $M \to N$ is defined to restrict to morphisms $M_j \to N_j$ for $j = 0, \ldots, k$. Similar notions, i.e., morphisms $\mathbb{M} \to \mathbb{N}$, make sense in terms of the respective stretched spaces. Now distributions on $\mathbb{M}$, say, in the meaning of elements of weighted Sobolev spaces on M (with k-tuples of multiple weights) are defined in terms of sums of localized contributions over $\mathbb{U}_{\text{sing}}$ and U, respectively.

5.3   Iterated Edge Spaces

Next we turn to an extension of the above Lemma (4.1), on how group actions on cone Sobolev spaces are motivated by reformulating edge spaces.

Lemma 5.1 *Let E be a Hilbert space with group action $\{\kappa_\lambda\}$. For every $s \in \mathbb{R}$ we have a canonical isomorphism*

$$I^s : \mathcal{W}^s(\mathbb{R}^{p+q}_{x,y}, E) \to \mathcal{W}^s(\mathbb{R}^p_x, \mathcal{W}^s(\mathbb{R}^q_y, E)), \tag{5.2}$$

induced by extending

$$I_{\mathcal{S}} : \mathcal{S}(\mathbb{R}^{p+q}_{x,y}, E) \to \mathcal{S}(\mathbb{R}^p_x) \hat{\otimes} \mathcal{S}(\mathbb{R}^q_y, E) \tag{5.3}$$

to (5.2), where the space $\mathcal{S}(\mathbb{R}^q_y, E)$ involved on the right-hand side of (5.3) is endowed with the group action $\{\chi_\lambda\}$ given by

$$(\chi_\lambda f)(y) = \kappa_\lambda \lambda^{q/2} f(\lambda y). \tag{5.4}$$

Proof Intuitively we will show, that E-valued distributions

$$f(x, y) \in \mathcal{W}^s(\mathbb{R}^{p+q}_{x,y}, E)$$

(functions in the Fourier image in variables (ξ, η)) can be interpreted as elements $f(y)$ taking values in x-dependent distributions, for $x \in \mathbb{R}^p$, and this is just the identification between corresponding spaces in (5.2), expressed in suitable norms, together with the right choice of (5.4). Since E-valued Schwartz functions are dense in the respective spaces it suffices to imagine that we consider those dense subspaces.

First, taking into account that

$$\|u\|_{\mathcal{W}^s(\mathbb{R}^q_y, E)} \sim \{\int (1 + |\eta|^2)^s \left\| \kappa^{-1}(\eta)(F_{y \to \eta} u)(\eta) \right\|^2_E d\eta\}^{1/2},$$

the left-hand side of (5.2) is defined (up to equivalence of norms) by

$$\left(\iint ([\xi]^2 + |\eta|^2)^s \left\| \kappa^{-1}(\xi, \eta)(F_{(x,y) \to (\xi,\eta)} f)(\xi, \eta) \right\|^2_E d\xi d\eta \right)^{\frac{1}{2}}, \tag{5.5}$$

where $f = f(x, y)$. Applying simple substitutions, the latter expression takes the form

$$\iint [\xi]^{2s} \left(1 + \frac{|\eta|^2}{[\xi]}\right)^s \left\| \kappa^{-1}(\xi, \eta) \hat{f}(\xi, \eta) \right\|_E^2 d\eta d\xi, \tag{5.6}$$

which is equal to

$$\int [\xi]^{2s} \int (1 + |\tilde{\eta}|^2)^s \left\| \kappa^{-1}(\xi, [\xi]\tilde{\eta}) \hat{f}(\xi, [\xi]\tilde{\eta}) \right\|_E^2 [\xi]^q d\tilde{\eta} d\xi. \tag{5.7}$$

On the other hand, for $\hat{f}(\xi, y) = (F_{x \to \xi} f)(\xi, y)$ by definition we have

$$\| f \|_{\mathcal{W}^s(\mathbb{R}_x^p, \mathcal{W}^s(\mathbb{R}_y^q, E))}^2 = \int [\xi]^{2s} \left\| \chi^{-1}(\xi) \hat{f}(\xi, y) \right\|_{\mathcal{W}^s(\mathbb{R}_y^q, E)}^2 d\xi \tag{5.8}$$

for $(\chi_\lambda u)(y) = \kappa_\lambda \lambda^{q/2} u(\lambda y)$. Clearly, χ_λ is a group action on the space $\mathcal{W}^s(\mathbb{R}_y^q, E)$; we easily see that $\chi_\lambda = \kappa_\lambda \lambda^\alpha \kappa_\lambda$, $\lambda \in \mathbb{R}_+$, for some real α. Thus (5.8) equals to

$$\int [\xi]^{2s} \left\| \kappa^{-1}(\xi)[\xi]^{-q/2} \hat{f}(\xi, [\xi]^{-1} y) \right\|_{\mathcal{W}^s(\mathbb{R}_y^q, E)}^2 d\xi \tag{5.9}$$

which is the same (up to equivalence of norms) as

$$\int [\xi]^{2s} \int (1 + |\eta|^2)^s \left\| \kappa^{-1}(\xi)\kappa^{-1}(\eta) \hat{\hat{f}}(\xi, \eta) \right\|_E^2 d\eta d\xi, \tag{5.10}$$

where $\kappa^{-1}(\xi)$ and $\kappa^{-1}(\eta)$ are acting in E. Remember that $\hat{\hat{f}}(\xi, \eta) := (F_{y \to \eta} \hat{f})(\xi, \eta) = (F_{(x,y) \to (\xi,\eta)} f)(\xi, \eta)$, and we employed relation (5.4) on how $\kappa^{-1}(\eta)$ commutes with the Fourier transform. Finally, the expressions (5.8) and (5.10) can be identified. In fact, substituting $\tilde{\xi} := \xi[\eta]$, expression (5.10) turns to

$$\int (1 + |\xi|^2)^s \int [\eta]^{2s} \left\| \kappa^{-1}(\xi)\kappa^{-1}(\eta) \hat{\hat{f}}(\xi, \eta) \right\|_E^2 d\eta d\xi \tag{5.11}$$

Using

$$[\xi, [\xi]\eta] \sim (\langle\xi\rangle^2 + \langle\xi\rangle^2 \langle\eta\rangle^2)^{\frac{1}{2}} \sim \langle\xi\rangle\langle\eta\rangle \sim [\xi][\eta] \tag{5.12}$$

with $\sim$ indicating estimates of the involved quantities up to constants, e.g.

$$c_1[\xi, [\xi]\eta] \le [\xi][\eta] \le c_2[\xi, [\xi]\eta]$$

with some positive constants c_1, c_2 for all $\xi, \eta \in \mathbb{R}^{p+q}$, in (5.7) we may replace $\kappa^{-1}(\xi, [\xi]\eta)$ by $\kappa^{-1}([\xi][\eta]) = \kappa^{-1}([\xi])\kappa^{-1}([\eta])$. Thus we see, that (5.7) is equivalent to (5.10). $\square$

Weighted Sobolev spaces on singular manifolds $\mathbb{M}$ are defined with the help of suitable partitions of unity, subordinate to an open covering like (5.1). If $M \in \mathfrak{M}_k$, then U is of singular order $k - 1$. and the contribution referring to U is already defined in the step before. Thus we may concentrate on $\mathbb{U}_{\mathrm{sing}}$. Then we choose a partition on unity subordinate to (5.1), called $(\omega_{\mathrm{sing}}, \omega_{\mathrm{reg}})$. When we denote the desired weighted edge spaces by $\mathcal{W}^{s,(\beta,\gamma)}(M)$ for tuples of weights $\beta = (\beta_1, \ldots, \beta_{k-1})$ and $\gamma \in \mathbb{R}$, then we set

$$\mathcal{W}^{s,(\beta,\gamma)}(M) = \omega_{\mathrm{sing}}\mathcal{W}^{s,(\beta,\gamma)}(U_{\mathrm{sing}}) + \omega_{\mathrm{reg}}\mathcal{W}^{s,\beta}(U). \tag{5.13}$$

Let us illustrate things in more detail for the case $M \in \mathfrak{M}_2$. First consider the case of an infinite stretched cone $B^\wedge$ for a compact manifold B with smooth edge Y. In this case we have $\beta, \gamma \in \mathbb{R}$.

Theorem 5.2 *Let* $A(\lambda) \in L^\mu(B, \mathbf{g}; \mathbb{R}^d_\lambda)$, $\mathbf{g} := (\beta, \beta - \mu)$; *then we have* $A^*(\lambda) \in L^\mu(B, \mathbf{g}^*; \mathbb{R}^d_\lambda)$, $\mathbf{g}^* := (-\beta + \mu, -\beta)$.

Theorem 5.3 ([75]) *Let H and $\tilde{H}$ be Hilbert spaces with group actions* $\kappa = \{\kappa_\iota\}_{\iota \in \mathbb{R}_+}$ *and* $\tilde{\kappa} = \{\tilde{\kappa}_\iota\}_{\iota \in \mathbb{R}_+}$, *respectively, and write*

$$\kappa(\eta) := \kappa_{\langle\eta\rangle}, \tilde{\kappa}(\eta) := \tilde{\kappa}_{\langle\eta\rangle}.$$

Assume that a function

$$a(y, \eta) \in C^\infty(\mathbb{R}^{2q}, \mathcal{L}(H, \tilde{H}))$$

satisfies the estimates

$$\pi(a) := \sup\big\{\|\tilde{\kappa}^{-1}(\eta)\{\partial_y^\alpha \partial_\eta^\beta a(y, \eta)\}\kappa(\eta)\|_{\mathcal{L}(H,\tilde{H})}$$
$$: (y, \eta) \in \mathbb{R}^{2q}, \alpha \leq \alpha_{\tilde{M}}, \beta \leq \beta_1\big\} < \infty, \tag{5.14}$$

where sup *is taken over all* $(y, \eta) \in \mathbb{R}^{2q}$ *and over* α, β *for* $\alpha_{\tilde{M}} := (\tilde{M} + 1, \ldots, \tilde{M} + 1)$, $\beta_1 := (1, \ldots, 1)$ *and* $\tilde{M} := M(\tilde{\kappa}) \in \mathbb{N}$ *associated with* $\tilde{\kappa}$. *Then* Op(a) *induces a continuous operator*

$$\mathrm{Op}(a) : \mathcal{W}^0(\mathbb{R}^q, H) \to \mathcal{W}^0(\mathbb{R}^q, \tilde{H}),$$

and we have

$$\|\mathrm{Op}(a)\|_{\mathcal{L}(\mathcal{W}^0(\mathbb{R}^q, H), \mathcal{W}^0(\mathbb{R}^q, \tilde{H}))} \le c\pi(a)$$

for some constant c independent of a.

$$t^{-\mu}r^{-\mu}\tilde{h}_1(t, w, r, tv, tr\eta, rt\lambda) \tag{5.15}$$

we can perform the calculus over B, while afterwards we may pass to a calculus that includes the variables t, w, and the parameter λ is multiplied by rt.

5.4   Distribution Spaces and Cone Operators

We now turn to the distribution spaces, playing a role in connection with $B^\wedge$. First we have spaces $\mathcal{K}^{s,\beta}(X^\wedge)$ for any real s and horizontal weight $\beta \in \mathbb{R}$. Remember, that β is the weight, locally belonging to $r \in \mathbb{R}_+$, and γ the vertical weight, belonging to $t \in \mathbb{R}_+$. We form

$$H_0 := \mathcal{H}^{s,\gamma}(Y^\wedge_{(y,t)}, \mathcal{K}^{s,\beta}(X^\wedge_{(x,r)})),$$

$$H_\infty := \mathcal{W}^s(Y^\wedge_{(y,t)}, \mathcal{K}^{s,\beta}(X^\wedge_{(x,r)})),$$

referring to charts $\mathbb{R}^{q+1}$ on $Y^\wedge$, a partition of unity and variables (y, t) on $Y^\wedge$, and

$$H_{int} := \mathcal{K}^{s,\gamma}((2\mathbb{B})^\wedge)|_{(B \setminus Y)^\wedge}.$$

Choose some cut-off function $\omega_1(t)$ which is equal to 1 near $t = 0$, and define

$$\mathcal{K}^{s,\beta,\gamma}(B^\wedge) := \sigma\{\omega_1(t)H_0 + (1 - \omega_1(t))H_\infty\} + (1 - \sigma)H_{int}, \tag{5.16}$$

where σ is a cut-off function with $\sigma = 1$ close to $Y^\wedge$. The space $\mathcal{K}^{s,\beta,\gamma}(B^\wedge)$ is an analogue of $\mathcal{K}^{s,\beta}(X^\wedge)$ in the calculus for a regular link. The space $\mathcal{K}^{s,\beta,\gamma}(B^\wedge)$ is endowed with a group action $\kappa_\iota : u(b, t) \mapsto \iota^{\frac{p+1}{2}}u(b, \iota t)$ for $p := \dim B$. We now turn to a more transparent description of spaces $\mathcal{K}^{s,\beta,\gamma}(B^\wedge)$, where $B \in \mathfrak{M}_1$ is a compact space with edge Y of dimension q. The associated double $2\mathbb{B}$ of the stretched space $\mathbb{B}$ belongs to $\mathfrak{M}_0$. Moreover, $\partial\mathbb{B}$ is a locally trivial bundle over Y with fiber $X \in \mathfrak{M}_0$. Choose an open covering of Y by coordinate neighborhoods

$$\mathcal{U} = \{U_1, \ldots, U_N\},$$

and corresponding charts $\chi_j : U_j \mapsto \mathbb{R}^q$, together with a subordinate partition of unity

$$\{\varphi_1, \ldots, \varphi_N\}, \varphi_j \in C_0^\infty(U_j).$$

We then have associated maps

$$(\chi_{1,j}^{-1})_* : \mathcal{W}^s(\mathbb{R}^q, \mathcal{K}^{s,\beta}(X^\wedge)) \mapsto \mathcal{W}^s(U_{1,j}, \mathcal{K}^{s,\beta}(X^\wedge)),$$

where the latter space is just a definition. Now we set

$$\mathcal{W}^s(\mathbb{R}_{+,t} \times U_j, \mathcal{K}^{s,\beta}(X_{(x,r)}^\wedge)) := (1 \times \chi_{1,j}^{-1})_* \mathcal{W}^s(\mathbb{R}^{1+q}, \mathcal{K}^{s,\beta}(X^\wedge))$$

and

$$\mathcal{W}_{cone}^s(Y_{(y,t)}^\wedge, \mathcal{K}^{s,\beta}(X_{(x,r)}^\wedge)) := \sum_{j=1}^{N}(1 \times \varphi_j)(1 \times \chi_{1,j}^{-1})_* \mathcal{W}^s(\mathbb{R}^{1+q}, \mathcal{K}^{s,\beta}(X^\wedge)).$$

In a similar manner we define

$$\mathcal{H}^{s,\gamma}(Y_{(y,t)}^\wedge, \mathcal{K}^{s,\beta}(X_{(x,r)}^\wedge)) := \sum_{j=1}^{N}(1 \times \varphi_j)(1 \times \chi_j^{-1})_* \mathcal{H}^{s,\gamma}(\mathbb{R}^{1+q}, \mathcal{K}^{s,\beta}(X^\wedge)).$$

This is just the detailed meaning of the spaces H_0 and H_∞, respectively, involved in $\mathcal{K}^{s,\beta,\gamma}(B^\wedge)$. What concerns the interior part H_{int} we first observe, that the double $2\mathbb{B}$ of the stretched space $\mathbb{B}$ is closed and compact. In fact, $\mathbb{B}$ is a manifold with smooth boundary $\partial\mathbb{B}$, consisting of a locally trivial $X^\wedge$-bundle over Y. Then $\partial\mathbb{B}^\wedge$ is an $X \times \mathbb{R}$-bundle over Y. Its restriction to $t = 0$ gives us an X-bundle over Y, which can be identified with $\partial\mathbb{B}$.

Bibliography

1. S. Agmon, A. Douglis, L. Nirenberg, Estimates near the boundary for solutions of elliptic partial differential equations satisfying general boundary conditions. I. Commun. Pure Appl. Math. **12**(4), 623–727 (1959). https://doi.org/10.1002/cpa.3160120405 (cited on page 104)
2. M.F. Atiyah, R. Bott, The index problem for manifolds with boundary, in *Differential Analysis: Papers Presented at the Bombay Colloquium, 1964*, ed. by M.F. Atiyah (Tata Institute of Fundamental Research/Oxford University Press, Bombay/Oxford, 1964), pp. 175–186 (cited on page vi)
3. M.F. Atiyah, R. Bott, The index problem for manifolds with boundary, in *Differential Analysis: Papers Presented at the International Colloquium*, ed. by M.S. Narasimhan. Tata Institute of Fundamental Research Studies in Mathematics, vol. 2. Proceedings of the International Colloquium, Bombay, 7–14 January 1964 (Oxford University Press, London, 1964), pp. 175–186. MR0185606, Zbl 0163.34603 (cited on page 104, 109)
4. M.F. Atiyah, I.M. Singer, The index of elliptic operators I, II, III. Ann. Math. **87**, 483–530, 531–545, 546–604 (1968) (cited on page v)
5. M.F. Atiyah, I.M. Singer, The index of elliptic operators. I. Ann. Math. **87**(3), 484–530 (1968). https://doi.org/10.2307/1970715 (cited on page 104)
6. M.F. Atiyah, V.K. Patodi, I.M. Singer, Spectral asymmetry and riemannian geometry. I. Math. Proc. Camb. Philos. Soc. **77**(1), 43–69 (1975). https://doi.org/10.1017/S0305004100049410 (cited on page 109)
7. M.F. Atiyah, V.K. Patodi, I.M. Singer, Spectral asymmetry and riemannian geometry. II. Math. Proc. Camb. Philos. Soc. **78**(3), 405–432 (1975). https://doi.org/10.1017/S0305004100051872 (cited on page 109)
8. M.F. Atiyah, V.K. Patodi, I.M. Singer, Spectral asymmetry and riemannian geometry. III. Math. Proc. Camb. Philos. Soc. **79**(1), 71–99 (1976). https://doi.org/10.1017/S0305004100052105 (cited on page 109)
9. L. Boutet de Monvel, Boundary problems for pseudo-differential operators. Acta. Math. **126**, 11–51 (1971). https://doi.org/10.1007/BF02392088 (cited on page vi, 4, 5, 45, 49, 64, 102, 103, 104, 106, 107, 109, 114, 118, 120)
10. A.P. Calderón, Singular integrals. Bull. Amer. Math. Soc. **72**(3), 427–465 (1966). https://doi.org/10.1090/S0002-9904-1966-11492-1 (cited on page 49)
11. D.-C. Chang, B.-W. Schulze, Calculus on spaces with higher singularities. J. Pseudo-Differ. Oper. Appl. **8**(4), 585–622 (2017). https://doi.org/10.1007/s11868-016-0180-x (cited on page 4)
12. D.-C. Chang, B.-W. Schulze, Ellipticity on spaces with higher singularities. Sci. China Math. **60**(11), 2053–2067 (2017). https://doi.org/10.1007/s11425-016-0519-9 (cited on page 4)

© Springer Nature Switzerland AG 2025

D.-C. Chang, B.-W. Schulze, *Analysis on Manifolds with Singularities*,
Pseudo-Differential Operators 16, https://doi.org/10.1007/978-3-032-01890-8

13. D.-C. Chang, A.J. Nagel, E.M. Stein, Estimates for the ∂-Neumann problem in pseudoconvex domains of finite type in C^2. Acta Math. **169**, 153–228 (1992). https://doi.org/10.1007/BF02392760 (cited on page 60)

14. D.-C. Chang, T. Qian, B.-W. Schulze, Corner boundary value problems. Complex Anal. Oper. Theory **9**(5), 1157–1210 (2014). https://doi.org/10.1007/s11785-014-0424-9 (cited on page 4)

15. D.-C. Chang, S. Khalil, B.-W. Schulze, Weighted corner spaces. AAO **3**(3), 391 (2019) (cited on page 65)

16. D.-C. Chang, S. Khalil, B.-W. Schulze, Corner operators with symbol hierarchies. Adv. Appl. Clifford Algebras **31**(3), 47 (2021). ISSN: 1661-4909. https://doi.org/10.1007/s00006-021-01130-x (cited on page 65)

17. J. Cheeger, On the spectral geometry of spaces with cone-like singularities. Proc. Nat. Acad. Sci. U.S.A. **76**, 2103–2106 (1979) (cited on page vii, 65)

18. H.O. Cordes, *The Technique of Pseudodifferential Operators*. London Mathematical Society Lecture Note Series, vol. 202 (Cambridge University Press, Cambridge, 1995). ISBN: 978-0-521-37864-2. https://doi.org/10.1017/CBO9780511569425 (cited on page 81, 123)

19. M. Dauge, *Elliptic Boundary Value Problems on Corner Domains*. Lecture Notes in Mathematics, vol. 1341 (Springer-Verlag, Berlin/Heidelberg, 1988) (cited on page vii)

20. N. Dines, Elliptic operators on corner manifolds. Ph.D. Thesis. University of Potsdam, 2006 (cited on page 83, 112)

21. C. Dorschfeldt. *Algebras of Pseudo-differential Operators Near Edge and Corner Singularities*. Mathematics Research, vol. 102 (Wiley-VCH, Berlin, 1998) (cited on page 112)

22. C. Dorschfeldt, B.-W. Schulze, Pseudo-differential operators with operator-valued symbols in the Mellin-edge-approach. Ann. Global Anal. Geom. **12**, 2 (1994) (cited on page 65)

23. J.V. Egorov, B.-W. Schulze, *Pseudo-differential Operators, Singularities, Applications*. Operator Theory: Advances and Applications, vol. 93 (Birkhäuser Verlag, Basel, 1997) (cited on page 5, 65)

24. G.I. Eskin, *Boundary Value Problems for Elliptic Pseudodifferential Equations*. Mathematical Monographs, vol. 52 (American Mathematical Society, Providence, 1980) (cited on page vi, vii, 5, 65, 102, 106, 118, 120)

25. H.-J. Flad, G. Harutyunyan, R. Schneider, B.-W. Schulze, Explicit Green operators for quantum mechanical Hamiltonians. I. Manuscripta Math. **135**, 497–519 (2011) (cited on page 5)

26. I.M. Gelfand, On elliptic equations. Uspechi Math. Nauk. **15**(3), 121–132 (1960). https://doi.org/10.1070/rm1960v015n03abeh004094 (cited on page v, 104)

27. J.B. Gil, B.-W. Schulze, J. Seiler, Holomorphic operator-valued symbols for edge-degenerate pseudo-differential operators. Differ. Equatiobs, Asymp. Anal. Math. Phys. Math. Res. **100**, 113–137 (1997) (cited on page 65, 112, 114, 115, 116)

28. J.B. Gil, B.-W. Schulze, J. Seiler, Cone pseudodifferential operators in the edge symbolic calculus. Osaka J. Math. **37**, 221–260 (2000) (cited on page 65, 84, 112, 114, 116, 117)

29. P.C. Greiner, E.M. Stein, *Estimates for the ∂-Neumann Problem*. Mathematical Notes, vol. 19 (Princeton University Press, Princeton, 1977). ISBN: 978-0-691-08013-0. https://doi.org/10.1515/9781400869220 (cited on page 60)

30. G. Grubb, *Functional Calculus of Pseudo-differential Boundary Problems*, 2nd edn. (Birkhäuser Verlag, Boston, 1996) (cited on page 64, 102, 120)

31. G. Harutyunyan, B.-W. Schulze, *Elliptic Mixed, Transmission and Singular Crack Problems* (European Mathematical Society, Zürich, 2008) (cited on page vi, 4, 45)

32. F. Hirzebruch, *Topological Methods in Algebraic Geometry*. Grundlehren der Mathematischen Wissenschaften, vol. 131 (Springer-Verlag, Berlin/Heidelberg/New York, 1966). https://doi.org/10.1007/978-3-662-30697-0 (cited on page 104)

33. L. Hörmander, Pseudo-differential operators. Commun. Pure Appl. Math. **18**, 3 (1965) (cited on page v, 4, 48)

34. L. Hörmander, *The Analysis of Linear Partial Differential Operators* (Springer-Verlag, New York, 1985) (cited on page 49, 102)

35. P. Jeanquartier, Transformation de Mellin et développements asymptotiques. Enseign. Math. **25**, 285–308 (1979) (cited on page 65)

36. D. Kapanadze, B.-W. Schulze, *Crack Theory and Edge Singularities* (Academic Publishing/Kluwer, Dordrecht, 2003) (cited on page vi, 4, 75)

37. S. Khalil, Boundary value problems on manifolds with singularities. Ph.D. Thesis. University of Potsdam, 2018 (cited on page 65)

38. J. Kohn, L. Nirenberg, An algebra of pseudo-differential operators. Commun. Pure Appl. Math. **18**, 269–305 (1965) (cited on page v, 4)

39. A.I. Komech, Elliptic boundary problems for pseudo-differential operators on manifolds with conical points. Mat. Sb **86**(2), 268–298 (1971) (cited on page vii)

40. V.A. Kondratyev, Boundary value problems for elliptic equations in domains with conical points. Trudy Mosk. Mat. Obshch. **16**, 209–292 (1967) (cited on page vii, 4, 65)

41. T. Krainer, Parabolic pseudodifferential operators and long-time asymptotics of solutions. Ph.D. Thesis. University of Potsdam, 2000 (cited on page 65, 83, 104, 112)

42. H. Kumano-Go, *Pseudo-Differential Operators* (The MIT Press, Cambridge/London, 1982). ISBN: 978-0-262-11080-8 (cited on page 13)

43. P.T. Lai, Problème de Dirichlet dans un cône avec paramètre spectral pour une classe d'espaces de Sobolev à poids. Commun. Part. Differ. Equ. **4**, 4 (1979) (cited on page vii)

44. X. Liu, B.-W. Schulze, *Boundary Value Problems with Global Projection Conditions*. Advances in Partial Differential Equations and Control, vol. 265 (Springer Nature Switzerland AG, Basel, 2018) (cited on page 104, 108, 109, 122)

45. X. Liu, B.-W. Schulze, *Boundary Value Problems with Global Projection Conditions*. Operator Theory: Advances and Applications, vol. 265 (Birkhäuser, Cham, 2018). https://doi.org/10.1007/978-3-319-70114-1 (cited on page 114)

46. R. Mazzeo, Elliptic theory of differential edge operators I. Commun. Part. Differ. Equ. **16**, 1615–1684 (1991) (cited on page 65)

47. R. Mazzeo, R.B. Melrose, *Pseudo-differential Operators on Manifolds with Fibred Boundaries*. Preprint (1998). arXiv:math./9812120 (cited on page 90)

48. R.B. Melrose, Transformation of bounday problems. Acta Math. **147**, 149–236 (1981) (cited on page 90)

49. R.B. Melrose, *Pseudo-differential Operators on Manifolds with Corners*. Manuscript (MIT, Boston, 1983) (cited on page vii)

50. R.B. Melrose, G. Mendoza, *Elliptic operators of totally characteristic type* (Mathematical Sciences Research Institute, Berkeley, 1983). MSRI Preprint, 047-83. https://www-math.mit.edu/~rbm/papers/b-elliptic-1983/1983-Melrose-Mendoza.pdf (cited on page 65)

51. C. Parenti, Operatori pseudo-differenziali in R^n e applicazioni. Ann. Mat. Pura Appl. **93**(1), 359–389 (1972). https://doi.org/10.1007/BF02412028 (cited on page 81)

52. A. Pietsch, Zur Theorie der topologischen Tensorprodukte. Math. Nachr. **25**, 19–30 (1963). https://doi.org/10.1002/mana.19630250104 (cited on page 109)

53. B.A. Plamenevskij, On the boundedness of singular integrals in spaces with weight. Mat. Sbornik **25**, 4 (1968) (cited on page vii, 65)

54. B.A. Plamenevskij, *Algebras of Pseudo-differential Operators* (Moscow, Nauka, 1986) (cited on page vii)

55. B.A. Plamenevskij, V.N. Senichkin, On the representation of algebras of pseudo-differential operators on manifolds with discontinuous symbols (Russian). Izv. Akad. Nauk SSSR **51**, 4 (1987) (cited on page vii)

56. V.S. Rabinovič, Pseudodifferential operators in non-bounded domains with conical structure at infinity. Mat. Sb. **80**, 77–97 (1969) (cited on page vii)

57. V.S. Rabinovich, Mellin pseudo-differential operators with operator symbols and their applications. Operator Theory: Advances and Applications, vol. 78 (Birkhäuser Verlag, Basel, 1995), pp. 271–279 (cited on page vii)

58. S. Rempel, B.-W. Schulze, Parametrices and boundary symbolic calculus for elliptic boundary problems without transmission property. Math. Nachr. **105**, 45–149 (1982) (cited on page vii, 4, 5)

59. S. Rempel, B.-W. Schulze. *Index Theory of Elliptic Boundary Problems*. Mathematische Monographien, vol. 55. (North Oxford Academic Publishing Company, Oxford, 1985). Transl. to Russian: Mir, Moscow. Akademie-Verlag, Berlin, 1982 (cited on page 5, 114, 120)

60. S. Rempel, B.-W. Schulze, Complete Mellin and Green symbolic calculus in spaces with conormal asymptotics. Ann. Glob. Anal. Geom. **4**, 2 (1986) (cited on page vii)

61. S. Rempel, B.-W. Schulze, Asymptotics for elliptic mixed boundary problems. Pseudo-differential and Mellin operators in spaces with conormal singularity. Math. Res. **50** (1989) (cited on page 65, 104, 109)

62. W. Rungrottheera, B.-W. Schulze, Weighted spaces on corner manifolds. Complex Var. Elliptic Equ. **59**(12), 1706–1738 (2014). ISSN: 1747-6933. https://doi.org/10.1080/17476933.2013.876416 (cited on page 4)

63. E. Schrohe, Spaces of weighted symbols and weighted sobolev spaces on manifolds, in *Pseudo-Differential Operators*, ed. by H. O. Cordes, B. Gramsch, H. Widom. Lecture Notes in Mathematics, vol. 1256 (Springer-Verlag, Berlin, 1987), pp. 360–377. https://doi.org/10.1007/BFb0077751 (cited on page 81)

64. E. Schrohe, B.-W. Schulze, Boundary value problems in Boutet de Monvel's calculus for manifolds with conical singularities I, in *Pseudo-Differential Calculus and Mathematical Physics*. Advance in Partial Differential Equations (Akademie Verlag, Berlin, 1994), pp. 97–209 (cited on page 104, 120)

65. E. Schrohe, B.-W. Schulze, Boundary value problems in Boutet de Monvel's calculus for manifolds with conical singularities II, in *Boundary Value Problems, Schrödinger Operators, Deformation Quantization*. Advance in Partial Differential Equations (Akademie Verlag, Berlin, 1995), pp. 70–205 (cited on page 90, 91)

66. B.-W. Schulze, Pseudo-differential operators on manifolds with edges. Teubner-Texte zur Mathematik **112**, 259–287 (1989) (cited on page 4)

67. B.-W. Schulze, *Pseudo-differential Operators on Manifolds with Singularities* (North-Holland, Amsterdam, 1991) (cited on page 112)

68. B.-W. Schulze, *Boundary Value Problems and Singular Pseudo-differential Operators* (John Wiley, Chichester, 1998) (cited on page 4, 5, 80, 87, 95, 108, 109, 112)

69. B.-W. Schulze, Operators with symbol hierarchies and iterated asymptotics. Publ. RIMS, Kyoto Univ. **38**, 4 (2002) (cited on page 4)

70. B.-W. Schulze, The iterative structure of the corner calculus, in *Pseudo-Differential Operators: Analysis, Application and Computations*, ed. by L. Rodino. Operator Theory: Advances and Applications (Birkhäuser Verlag, Basel, 2011), pp. 79–103 (cited on page vi, 5)

71. B.-W. Schulze, M.W. Wong, Mellin and Green operators of the corner calculus. J. Pseudo-Differ. Oper. Appl. **2**, 4 (2011) (cited on page 4)

72. B.-W. Schulze, J. Seiler, Truncation quantization in the edge calculus. J. Pseudo-Differ. Oper. Appl. **14**(4), 69 (2023). ISSN: 1662-999X. https://doi.org/10.1007/s11868-023-00564-0 (cited on page 64, 65, 105)

73. B.-W. Schulze, Y. Wei, The Mellin-Edge quantisation for corner operators. Complex Anal. Oper. Theory **8**(4), 803–841 (2014). ISSN: 1661-8262. https://doi.org/10.1007/s11785-013-0289-3 (cited on page 4)

74. J. Seiler, Pseudodifferential calculus on manifolds with non-compact edges. Ph.D. Thesis. University of Potsdam, 1997 (cited on page 93)

75. J. Seiler, Continuity of edge and corner pseudo-differential operators. Math. Nachr. **205**, 163–182 (1999) (cited on page 4, 127)

76. J. Seiler, Calculus for parametric boundary problems with global projection conditions. J. Funct. Anal. **289**(10), 111099 (2025). https://doi.org/10.1016/j.jfa.2025.111099 (cited on page 109)

77. J. Seiler, *Calculus for Parametric Boundary Problems with Global Projection Conditions*. Preprint (2024). arXiv:2410.12615. https://doi.org/10.48550/arXiv.2410.12615 (cited on page 109)

78. M.A. Shubin, Pseudodifferential operators in R^n. Doklady Akademii Nauk SSSR **196**(2), 316–319 (1971) (cited on page 81)

79. M.A. Shubin, *Pseudodifferential Operators and Spectral Theory* (Berlin, Springer-Verlag, 1987) (cited on page 4)

80. E.M. Stein, *Harmonic Analysis: Real-Variable Methods, Orthogonality, and Oscillatory Integrals*. Princeton Mathematical Series, vol. 43 (Princeton University Press, Princeton, 1993). ISBN: 978-0-691-03216-0. https://doi.org/10.1515/9781400883929 (cited on page 54, 58)

81. M.I. Vishik, G.I. Eskin, Convolution equations in a bounded region. Uspekhi Mat. Nauk **20**, 3 (1965) (cited on page vi)

82. M.I. Vishik, G.I. Eskin, Convolution equations in bounded domains in spaces with weighted norms. Mat. Sb. **69**, 1 (1966) (cited on page vi)

83. I. Witt, On the factorization of meromorphic Mellin symbols, in *Parabolicity, Volterra Calculus, and Conical Singularities*. Operator Theory, Advances and Applications, vol. 138 (Birkhäuser Verlag, Basel, 2002), pp. 279–306 (cited on page 65)

MIX
Papier aus verantwortungsvollen Quellen
Paper from responsible sources
FSC® C105338

If you have any concerns about our products,
you can contact us on
ProductSafety@springernature.com

In case Publisher is established outside the EU,
the EU authorized representative is:
Springer Nature Customer Service Center GmbH
Europaplatz 3, 69115 Heidelberg, Germany

Printed by Libri Plureos GmbH
in Hamburg, Germany